Solving Statics Problems in Maple

Brian D. Harper
Ohio State University

ENGINEERING MECHANICS
STATICS

Sixth Edition

J. L. Meriam
L. G. Kraige
*Virginia Polytechnic Institute
and State University*

John Wiley & Sons, Inc.

Cover Photo: © Medioimages/Media Bakery

To order books or for customer service please, call 1-800-CALL WILEY (225-5945).

ISBN-13   978- 0-470-09923-0
ISBN-10   0-470-09923-2

10 9 8 7 6 5 4 3 2 1

CONTENTS

INTRODUCTION

Computers and software have had a tremendous impact upon engineering education over the past several years and most engineering schools now incorporate computational software such as Maple in their curriculum. Since you have this supplement the chances are pretty good that you are already aware of this and will be having to learn to use Maple as part of a Statics course. The purpose of this supplement is to help you do just that.

There seems to be some disagreement among engineering educators regarding how computers should be used in an engineering course such as Statics. I will use this as an opportunity to give my own philosophy along with a little advice. In trying to master the fundamentals of Statics there is no substitute hard work. The old fashioned taking of pencil to paper, drawing free body diagrams and struggling with equilibrium equations, etc. is still essential to grasping the fundamentals of mechanics. A sophisticated computational program is not going to help you to understand the fundamentals. For this reason, my advice is to use the computer only when required to do so. Most of your homework can and should be done without a computer. A possible exception might be using Maple's symbolic algebra capabilities to check some messy calculations.

The problems in this booklet are based upon problems taken from your text. The problems are slightly modified since most of the problems in your book do not require a computer for the reasons discussed in the last paragraph. One of the most important uses of the computer in studying Mechanics is the convenience and relative simplicity of conducting parametric studies. A parametric study seeks to understand the effect of one or more variables (parameters) upon a general solution. This is in contrast to a typical homework problem where you generally want to find one solution to a problem under some specified conditions. For example, in a typical homework problem you might be asked to find the reactions at the supports of a structure with a concentrated force of magnitude 200 lb that is oriented at an angle of 30 degrees from the horizontal. In a parametric study of the same problem you might typically find the reactions as a function of two parameters, the magnitude of the force and its orientation. You might then be asked to plot the reactions as a function of the magnitude of the force for several different orientations. A plot of this type is very beneficial in visualizing the general solution to a problem over a broad range of variables as opposed to a single case.

As you will see, it is not uncommon to find Mechanics problems that yield equations that cannot be solved exactly. These problems require a numerical approach that is greatly simplified by computational software such as Maple. Although numerical solutions are extremely easy to obtain in Maple this is still the method of last resort. Chapter 1 will illustrate several methods for obtaining symbolic (exact) solutions to problems. These methods should always be tried first. Only when these fail should you generate a numerical approximation.

Many students encounter some difficulties the first time they try to use a computer as an aid to solving a problem. In many cases they are expecting that they have to do something fundamentally different. It is very important to understand that there is no fundamental difference in the way that you would formulate computer problems as opposed to a regular homework problem. Each problem in this booklet has a problem formulation section prior to the solution. As you work through the problems be sure to note that there is nothing peculiar about the way the problems are formulated. You will see free-body diagrams, equilibrium equations etc. just like you would normally write. The main difference is that most of the problems will be parametric studies as discussed above. In a parametric study you will have at least one and possibly more parameters or variables that are left undefined during the formulation. For example, you might have a general angle θ as opposed to a specific angle of $20°$. If it helps, you can "pretend" that the variable is some specific number while you are formulating a problem.

This supplement has seven chapters. The first chapter contains a brief introduction to Maple. If you already have some familiarity with Maple you can skip this chapter. Although the first chapter is relatively brief it does introduce all the methods that will be used later in the book and assumes no prior knowledge of Maple. Chapters 2 through 7 contain computer problems taken from chapters 2 through 7 of your textbook. Thus, if you would like to see some computer problems related to friction you can look at the problems in chapter 6 of this supplement. Each chapter will have a short introduction followed by a table which summarizes the main commands and methods used along with the problem in which they are used. This would be the ideal place to look if you are interested in find examples of how to use specific functions, operations etc.

AN INTRODUCTION TO MAPLE

1

This chapter provides an introduction to the Maple programming language. Although the chapter is introductory in nature it will cover everything needed to solve the computer problems in this booklet.

1.1 Numerical and Symbolic Calculations

A command line in a Maple worksheet contains a set of instructions followed by a semicolon or colon. Pressing enter causes Maple to process the line. If the line ends in a semicolon Maple will show the output on the next line. If the line ends in a colon Maple will still process the line but the output will not be shown. It is a good idea to always use a semicolon when you are writing and debugging a worksheet. Later, if you wish, you can change the line to end with a colon to hide superfluous or lengthy output. There are several places in this supplement where the output will be hidden by using a colon. This is done either for trivial assignments where output is not necessary or for those rare occasions where the output may be several lines or even pages.

In an actual Maple session, command lines will begin with "[>" and will be red. Output is centered and blue. In this supplement, command lines will begin only with ">" while output lines will follow on the next line and be indented. The order in which command lines are executed is important in Maple, however, the order in which they are written is not. For this reason you can execute a line once and then go back to the same line later and execute it again, perhaps getting a different result. Thus, it is almost inevitable that something will go wrong as you debug a program. If you get some unexpected result the first thing you should try is re-executing your worksheet. To this end it is highly advisable to start every session with the *restart* command. Then, if you need to have the program re-execute from the beginning you can select *Edit...Execute...Worksheet*.

Here are several examples illustrating typical numerical calculations.

```
> restart; # always start with restart
> 2+7-4;
```

$$5$$

```
> 5*12-4*3;
```
$$48$$

```
> 8^2;
```
$$64$$

```
> 3/10+12/4;
```
$$\frac{33}{10}$$

```
> sqrt(2);
```
$$\sqrt{2}$$

Note in the last two examples that Maple will not show a decimal representation of a number unless asked to do so. Converting a number to a decimal approximation is accomplished with the *evalf* function.

```
> evalf(sqrt(2));
```
$$1.414213562$$

By default, the number of significant digits is 10. You can specify the number of significant digits in the *evalf* command as follows.

```
> evalf(sqrt(2),5);
```
$$1.4142$$

You can also change the default with the *Digits* variable.

```
> Digits:=20:
> evalf(sqrt(2));
```
$$1.4142135623730950488$$

Later we will find that the *evalf* function can be used as an easy way to perform numerical integration.

In the previous examples we have been using Maple as if it were an expensive calculator. Maple is, however, much more than a calculator. It can, for example, perform algebraic manipulations upon variables as well as numbers. Such manipulations are generally referred to as symbolic calculations or computer algebra. Following are several simple examples of symbolic calculations.

```
> (a*x+b*x^2)/x^3;
```
$$\frac{a\,x + b\,x^2}{x^3}$$

```
> %*x^2;
```

$$\frac{a\,x + b\,x^2}{x}$$

Note that % is a shortcut for referring to the last expression evaluated by Maple. Similarly, %% would refer to the next to last expression and %%% to the one before that. Be sure you understand the following calculation.

> %*%%;

$$\frac{(a\,x + b\,x^2)^2}{x^4}$$

Sometimes Maple does not produce the simplest form of a symbolic operation. If you suspect this might be the case, try simplifying with the *simplify* command.

> simplify(%);

$$\frac{(a + b\,x)^2}{x^2}$$

Here is another example of the *simplify* command.

> (x^2+x^4)/(x^3);

$$\frac{x^2 + x^4}{x^3}$$

> simplify((x^2+x^4)/(x^3));

$$\frac{1 + x^2}{x}$$

1.2 Assignments, Names and Variables

The assignment command ":=" (two keystrokes, colon followed by equal sign) can be used to assign an object (usually a number or an expression) to a variable.

> restart;
> x := 3;

$$x := 3$$

> We := Love +Statics;

$$We := Love + Statics$$

In the first example, the number 3 is assigned to the variable x. In the second, the sum of two variables is assigned to a third variable We. Note that Maple is case sensitive. *We, we, wE and WE* are all different variables. An assigned variable is sometimes referred to as a name or an alias. Whenever Maple encounters *We* it will automatically substitute *Love + Statics*. Thus, it is also useful to think of assignments (names, aliases) as abbreviations. "*We*" is an abbreviation for "*Love + Statics*". To illustrate this feature, consider the following symbolic calculations.

> We^2;

$$(Love + Statics)^2$$

> 10^We;

$$10^{(Love + Statics)}$$

> We/We;

$$1$$

> Love:=9: Statics:= 16:
> sqrt(We);

$$5$$

It is important to understand the difference between the assignment operator (:=) and the equality operator (=). := assigns an object to a variable while = defines equalities. We'll use the equality operator primarily to write equations that we subsequently ask Maple to solve. This will be discussed in greater detail later. Here we will just give a simple example containing both operators.

> eqn := x^2 + y^2 = z^2;

$$eqn := 9 + y^2 = z^2$$

Here we have assigned an equality to the variable *eqn*. You may recall that *x* has been previously assigned the value 3. Maple remembers this and automatically substitutes 3 for *x*. The *unassign* command removes a previous assignment.

> x;

$$3$$

> unassign('x'); # note that the variable must be placed in quotations.
> x;

$$x$$

Listing x before and after the *unassign* command just reinforces the change in assignment and is not necessary. It is important to understand at this point that the assignment for *eqn* has not changed even though the variable *x* is no longer assigned. To see this let's list the variable *eqn*.

> eqn;

$$9 + y^2 = z^2$$

The reason for this is that the variable *eqn* was assigned when *x* was equal to 3. Thus, unassigning *x* has no effect on that previous assignment. If you want the name *eqn* to contain the variable *x* then *x* should be unassigned before *eqn* is assigned. This point is illustrated by the following.

```
> eqn := x^2 + y^2 = z^2;
```
$$eqn := x^2 + y^2 = z^2$$

```
> x:=3:
> eqn;
```
$$9 + y^2 = z^2$$

```
> unassign('x');
> eqn;
```
$$x^2 + y^2 = z^2$$

Now we see that unassigning x recovers the original definition of eqn.

1.3 Functions

Maple has many basic mathematical functions built in. The following table provides a summary of those most useful in Statics.

Table 1.1 Built in mathematical functions.

sin, cos, tan, cot, sec, etc.	Trig functions, Sine, Cosine etc.
arcsin, arccos, arctan, arcsec, etc.	Inverse trig functions
^	Raising to a power, $x^\wedge y = x^y$.
exp	Exponential function. $\exp(y) = e^y$.
log	Natural logarithm.
Log10	Base 10 logarithm.
sqrt	Square root, $sqrt(x) = \sqrt{x}$

When using trig functions, be sure to remember that Maple uses radians by default. Also, the number π is represented in Maple by Pi while pi is just the greek letter π. As the following lines show, it is not always easy to tell one from the other in Maple output.

```
> restart;
> Pi; pi;
```
$$\pi$$

$$\pi$$

Using Pi and pi as arguments in a function makes it clear that the first is a number while the second is a variable.

> sin(Pi/2); sin(pi/2);
                1

$$\sin\left(\frac{1}{2}\pi\right)$$

All the greek letters (with the exception of the protected names Pi and gamma) can be used by spelling out their names. Type ?greek at a command prompt for a review of the greek letters and their spelling. Here are a few examples.

> cos(theta)+sin(Theta); tan(omega)-cot(Omega);
          $\cos(\theta) + \sin(\Theta)$

          $\tan(\omega) - \cot(\Omega)$

It turns out that the greek letter gamma (γ) is commonly used in Statics problems to denote an angle. If you do not need the gamma function (and you probably won't), feel free to unprotect it and then use it in your problems.

> gamma:=Pi/4;
Error, attempting to assign to `gamma` which is protected

> unprotect('gamma');
> gamma:=Pi/4;
          $$\gamma := \frac{1}{4}\pi$$

Maple also allows the user to define functions. Here are two examples.

> f:= x ->x^3;
          $f := x \rightarrow x^3$

> g:= (x,y) -> x^2+y^2;
          $g := (x,y) \rightarrow x^2 + y^2$

The syntax for defining a function starts with the usual assignment operator ":=" followed by a list of variables and then what looks like an arrow but is really two key strokes, a minus sign "-" and a greater than symbol ">". Note that more than one variable requires the use of parentheses. User defined functions operate exactly like built in functions. Their arguments may be numbers, variables or expressions.

> f(3);
          27

> f(2)+g(1,2)/g(3,1);
          $$\frac{17}{2}$$

> f(g(2,4));
$$8000$$

> g(duck,goose);
$$duck^2 + goose^2$$

> f(g(r^2,sqrt(r)));
$$\left(r^4 + r\right)^3$$

It is important to understand that there is nothing special about the choice of variables used to define a function. The following lines show that the structure of the function is not changed even if x and y are assigned.

> x:=3: y:=3:
> g(x,y);
$$18$$

> g(m,q);
$$m^2 + q^2$$

1.4 Graphics

One of the most useful things about a computational software package such as Maple is the ability to easily create graphs of functions. As we will see, these graphs allow one to gain a lot of insight into a problem by observing how a solution changes as some parameter (the magnitude of a load, an angle, a dimension etc.) is varied. This is so important that practically every problem in this supplement will contain at least one plot. By the time you have finished reading this supplement you should be very proficient at plotting in Maple. This section will introduce you to the basics of plotting in Maple.

Simple Plots of One or More Expressions

Simple plots are easily obtained with the *plot* command. The simplest plot call requires only the expression to be plotted and the range over which it is to be plotted. There are many advanced features of the plot command allowing graphs to be formatted in various ways. We will show the main options primarily by way of example. For more details, type ?plot, ?plot[structure], or ?plot[options] at any Maple command prompt ([>).

> restart; # always start with restart
> plot(5*x^2-0.5*x^3, x=0..10, color=black);

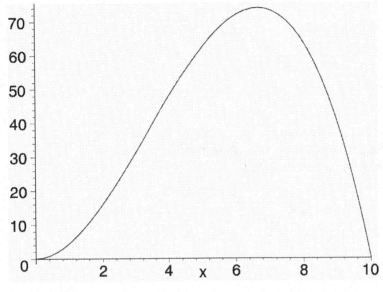

```
> f:=x^2*sin(x);
```
$$f := x^2 \sin(x)$$

```
> plot(f, x=-2*Pi..2*Pi, title=`x^2sin(x)`);
```

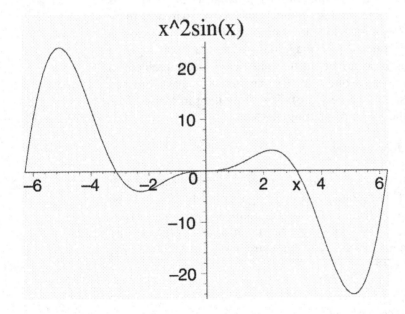

Plotting more than one function on a single graph is accomplished by replacing the single function with a list of functions in brackets [].

> g:=sqrt(x)*exp(-x); h:=x^2*10^(-x/2);

$$g := \sqrt{x}\; e^{(-x)}$$

$$h := x^2\, 10^{(-1/2\,x)}$$

> plot([g,h],x=0..5, color=black);

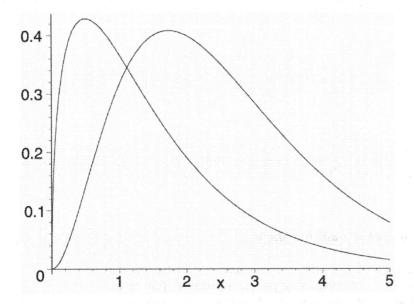

In an actual Maple session the two curves above would be plotted in different colors. If you want to distinguish results by color, its a good idea to use the color option. For example, the plot command plot([g,h],x=0..5, color=[red,blue]) would yield a graph where g is plotted red and h blue.

It is also possible, though somewhat tedious, to label individual curves in a plot. This requires the *display* and *textplot* procedures which must be loaded from the plots package with the command with(plots).

> with(plots):
> p1:=plot([g,h],x=0..5, color=black):
> t1:=textplot([1,0.41,"g"],color=black,font=[TIMES,ROMAN,16]):
> t2:=textplot([2.3,0.41,"h"],color=black,font=[TIMES,ROMAN,16]):

The parameters for *textplot* are the text and the coordinates at which it is to be plotted. Note that the individual *plot* and *textplot* structures are assigned names. These names are then input to the *display* procedure.

> display(p1,t1,t2);

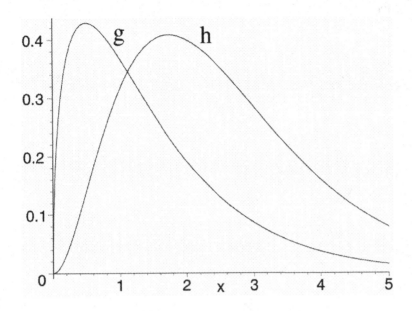

Multiple Plots of a Single Function (Parametric Studies)

One of the most important uses of the computer in studying Statics is the convenience and relative simplicity of conducting parametric studies (not to be confused with parametric plotting). A parametric study seeks to understand the effect of one or more variables (parameters) upon a general solution. This is in contrast to a typical homework problem where you generally want to find one solution to a problem under some specified conditions. For example, in a typical homework problem you might be asked to find the reactions at the supports of a structure with a concentrated force of magnitude 200 lb that is oriented at an angle of 30 degrees from the horizontal. In a parametric study of the same problem you might typically find the reactions as a function of two parameters, the magnitude of the force and its orientation. You might then be asked to plot the reactions as a function of the magnitude of the force for several different orientations. A plot of this type is very beneficial in visualizing the general solution to a problem over a broad range of variables as opposed to a single case.

Parametric studies generally require making multiple plots of the same function with different values of a particular parameter in the function. Following is a very simple example.

```
> restart;
> f:=5+x-5*x^2+a*x^3;
```
$$f := 5 + x - 5 x^2 + a x^3$$

What we would like to do is gain some understanding of how f varies with both x and a. It might be tempting to make a three dimensional plot in a case like this. Such a plot can, in some cases, be very useful. Usually, however, it is too difficult to interpret. This is illustrated by the following three dimensional plot of f versus x and a.

> plot3d(f,x=-10..10,a=-3..3,color=grey);

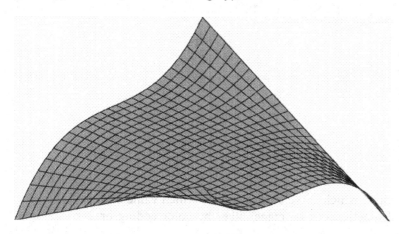

The plot above certainly is interesting but, as mentioned above, not very easy to interpret. In most cases it is much better to plot the function several times (with different values of the parameter of interest) on a single two dimensional graph. We will illustrate this by plotting f as a function of x for four values of the parameter a. There are a number of ways this can be done. Below are two methods: (a) re-assignment and (b) substitution.

> a:=-2: f1:=f: a:=-1: f2:=f: a:=1: f3:=f: a:=2: f4:=f: # method (a)
> plot([f1,f2,f3,f4],x=-10..10,color=black);

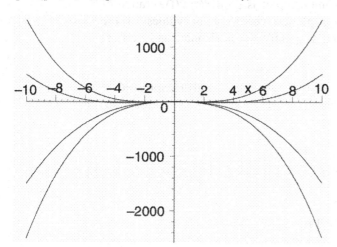

```
> unassign('a');
> plot([subs(a=-2,f),subs(a=-1,f),subs(a=1,f),subs(a=2,f)],x=-10..10,color=black); # method (b)
```

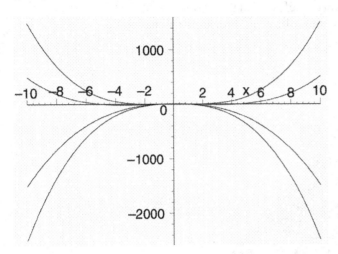

It is almost impossible to determine which curve corresponds to which value of a in the above two plots. This distinction can be made either by color coding or labeling, as discussed earlier.

Parametric Plots

It often happens that one needs to plot some function y versus x but y is not known explicitly as a function of x. For example, suppose you know the x and y coordinates of a particle as a function of time but want to plot the trajectory of the particle, i.e. you want to plot the y coordinate of the particle versus the x coordinate. Since both x and y are known in terms of another parameter (in this case time t), it is possible to obtain the desired plot using a parametric plot in Maple. The calling sequence for the parametric plot is plot([x(t),y(t),t=range of t],h,v,options), which results in y being plotted versus x for values of the parameter t in the range specified. h and v specify the horizontal and vertical ranges for the plot.

Now let's consider a simple example with two functions f and g expressed in terms of a parameter t.

```
> f:=10*t*(2-t); g:=sin(5*t);
```
$$f := 10\, t\,(2-t)$$

$$g := \sin(5\, t)$$

To plot f versus g:
```
> plot([g, f, t=0..2],color=black,labels=['g','f']);
```

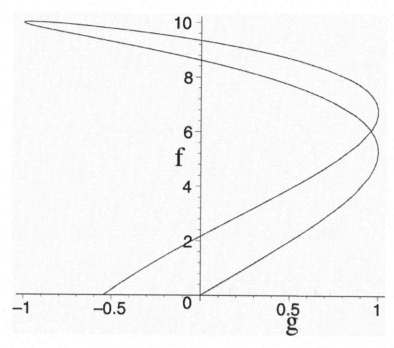

To plot several parametric plots on the same graph one should use the display procedure. Recall that this procedure must be loaded by typing with(plots): As an example, consider the following expressions for x and y in terms of a common parameter t:

```
> with(plots):
> y:=a*t-b*t^2;x:=c*sin(beta*t);
```

$$y := a t - b t^2$$

$$x := c \sin(\beta t)$$

Now suppose we want to specify that a=1, b=0.5, c=2 and then plot y versus x for three values of β (0.1, 0.25, 0.5). This is accomplished as follows.

```
> a:=1: b:=0.5: c:=2:
> beta:=0.1:p1:=plot([x, y, t=0..2],0..2,0..0.6,color=black):
> beta:=0.25:p2:=plot([x, y, t=0..2],0..2,0..0.6,color=black):
> beta:=0.5:p3:=plot([x, y, t=0..2],0..2,0..0.6,color=black):
> display([p1,p2,p3]);
```

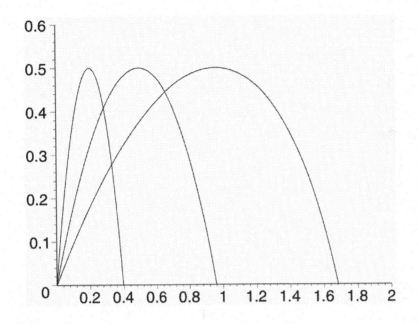

1.5 Differentiation and Integration

Mechanics problems often require integration and differentiation. In Maple, you can perform these operations either numerically or symbolically. In this supplement we will generally find symbolic results whenever possible and leave numerical solutions as a method of last resort.

```
> restart;
> diff(x*tan(x^3),x);
```
$$\tan(x^3) + 3\,x^3\,(1 + \tan(x^3)^2)$$

```
> f:=a*sec(b*t);
```
$$f := a \sec(b\,t)$$

```
> diff(f,t);
```
$$a \sec(b\,t)\,\tan(b\,t)\,b$$

Higher order derivatives can be obtained in two ways as illustrated by the following.

```
> g:=a*log(b*x^2);
```
$$g := a \ln(b\,x^2)$$

```
> diff(g,x,x,x); diff(g,x$3); # each expression evaluates the third derivative of g.
```

$$4\frac{a}{x^3}$$

$$4\frac{a}{x^3}$$

You can also use derivatives in defining functions. As an example, suppose a particle moves in a straight line and its position s is known as a function of time. From your elementary physics course you probably know that the first and second derivatives of the position give the velocity and acceleration of the particle.

> s:=10*t-20*t^2+2*t^3;
$$s := 10\,t - 20\,t^2 + 2\,t^3$$

> v:=diff(s, t);
$$v := 10 - 40\,t + 6\,t^2$$

> a:=diff(s, t$2);
$$a := -40 + 12\,t$$

Symbolic integration will be performed with the *int* command. The general format of this command is int(f, x = a..b) where f is the integrand (a Maple expression), x is the integration variable and a and b are the limits of integration. If the integration limits are omitted, the indefinite integral will be evaluated. Here are several examples of definite and indefinite integrals.

> restart;
> int(sin(b*x),x);
$$-\frac{\cos(b\,x)}{b}$$

> int(sin(b*x),x = a..b);
$$\frac{-\cos(b^2) + \cos(b\,a)}{b}$$

> g:=b*log(x);
$$g := b\ln(x)$$

> int(g,x);
$$b\,x\ln(x) - b\,x$$

> int(g,x = c..d);
$$b\,d\ln(d) - b\,d - b\,c\ln(c) + b\,c$$

If a definite integral contains no unknown parameters either in the integrand or the integration limits, the *int* command will provide numerical answers. Here are a few examples.

> int(x+3*x^3, x = 0..3);
$$\frac{261}{4}$$

> int(log(x), x = 2..5); # don't forget that log is the natural logarithm
$$-3 + 5 \ln(5) - 2 \ln(2)$$

Note that Maple will always try to return an exact answer. This usually results in answers containing fractions or functions as in the above examples. This is very useful in some situations; however, one often wants to know the numerical answer without having to evaluate a result such as the above with a calculator. To obtain numerical answers use Maple's *evalf* function as in the following examples.

> evalf(int(x+3*x^3, x = 0..3));
$$65.25000000$$

> evalf(int(log(x), x = 2..5));
$$3.660895199$$

1.6 Solving Equations

Solving one or more equations in Maple is usually accomplished with the solve command. The calling sequence for the solve command is solve(eqns, vars) where "eqns" is a set of equations and "vars" is a set of variables to solve for. There may be more unknowns in a set of equations than appear in the "vars" list. If the number of equations equals the number of unknowns, Maple will return a numerical answer. If the number of equations is less than the number of unknowns, Maple will evaluate the expression symbolically.

Solving Single Equations

Let's consider the simple case of finding the roots of a quadratic equation, $a x^2 + b x + c = 0$.

> restart;
> solve(a*x^2+b*x+c = 0,x);
$$-\frac{b - \sqrt{b^2 - 4ac}}{2a}, -\frac{b + \sqrt{b^2 - 4ac}}{2a}$$

Here is another example.

> v:=3-6*t+2*t^2;
$$v := 3 - 6t + 2t^2$$

We have a general expression for v as a function of t. We can easily find the value of t at which v = 2 as follows.

> solve(v=2,t);
$$\frac{3}{2}+\frac{1}{2}\sqrt{7}, \frac{3}{2}-\frac{1}{2}\sqrt{7}$$

If we don't want to get our calculators out to evaluate the above expressions for t, we should use the following:

> evalf(solve(v=2,t));
$$2.822875656 , .177124344$$

We can also use the solve command to get a general expression for t given any specified v = v0.

> solve(v=v0,t);
$$\frac{3}{2}+\frac{1}{2}\sqrt{3+2\,v0}, \frac{3}{2}-\frac{1}{2}\sqrt{3+2\,v0}$$

As another example, assume v is a function of an angle θ and we would like to find the values of θ at which v has a specific value:

> v:=4*sin(2*theta)-2*cos(2*theta);
$$v := 4\sin(2\,\theta) - 2\cos(2\,\theta)$$

> solve(v=v0,theta);
$$\frac{1}{2}\arctan\left(\frac{v0}{5}+\frac{\sqrt{-v0^2+20}}{10}, -\frac{v0}{10}+\frac{\sqrt{-v0^2+20}}{5}\right),$$
$$\frac{1}{2}\arctan\left(\frac{v0}{5}-\frac{\sqrt{-v0^2+20}}{10}, -\frac{v0}{10}-\frac{\sqrt{-v0^2+20}}{5}\right)$$

Occasionally when solving equations in Maple you may encounter the following.

> restart;
> f:=10*sin(x)=2*x^2+1;
$$f := 10\sin(x) = 2\,x^2 + 1$$

> solve(f,x);
$$\text{RootOf}(-10\sin(\_Z) + 2\,\_Z^2 + 1, label=\_L1)$$

It is beyond the scope of this chapter to get into exactly what the *RootOf* result means in Maple. Suffice it to say that Maple was not able to find an exact (symbolic) solution of the equation. There are two general approaches to obtaining an approximate solution that you might consider in a case like this; graphical and numerical.

With a graphical approach you could consider making a plot of the expressions on either side of the equals sign and then finding where the two curves intersect.

> plot([10*sin(x), 2*x^2+1], x=0...3);

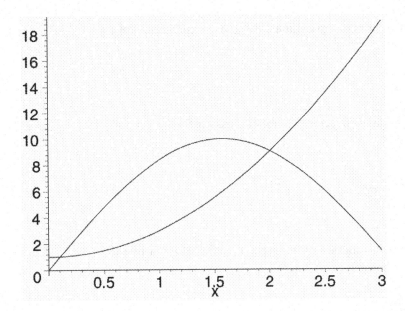

From the plot above we see that there are two solutions at x equals about 0.1 and 2.0. Of course, you might have to experiment around with the plot limits to make sure that you have identified all the solutions. To obtain a more accurate answer you should print the graph and then draw vertical lines at the intersections of the two curves. The intersections of these lines with the x-axis will give the two solutions. Another approach, which avoids having to draw vertical lines, is to rewrite the equation so that you have zero on the right hand side.

> g:=10*sin(x)-2*x^2-1;
$$g := 10\sin(x) - 2x^2 - 1$$

Observe that the values of x for which g = 0 will be solutions to our original equation. Now plot g versus x and find where the curve intersects the x axis.

> plot(g, x=0..2.5);

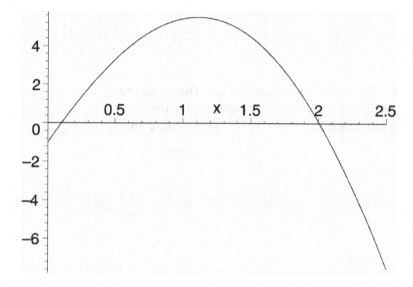

A much faster approach is to use fsolve to obtain a numerical solution.

> fsolve(f,x);
                .1022700143

This was certainly quick, but note also a danger of missing multiple solutions. Thus it is generally a good idea to make a plot even when you are going to use *fsolve*. From the plot we can pick up the second solution by using fsolve and specifying a range.

> fsolve(f,x,x=1..3);
                2.007613807

Finding Maxima and Minima of Functions

The usual method for finding maxima or minima of a function f(x) is to first determine the location(s) x at which maxima or minima occur by solving the equation $\dfrac{df}{dx}=0$ for x. One then substitutes the value(s) of x thus determined into f(x) to find the maximum or minimum. Consider finding the maximum velocity given the following expression for the velocity as a function of time:

> v:=b+c*t-d*t^3;
                $$v := b + c\,t - d\,t^3$$

> tm:=solve(diff(v,t)=0,t);
                $$tm := \frac{\sqrt{3}\,\sqrt{d\,c}}{3\,d}, \; -\frac{\sqrt{3}\,\sqrt{d\,c}}{3\,d}$$

> vm:=subs(t=tm[1],v);

$$vm := b + \frac{c\sqrt{3}\sqrt{dc}}{3\,d} - \frac{\sqrt{3}\,(dc)^{(3/2)}}{9\,d^2}$$

Note that t = tm[1] substitutes the first of the two results for tm. The second root is negative and thus not physically significant. Whether the result vm corresponds to a minimum or a maximum depends on the values of b, c and d. Let's look at a specific case:

> b:=1: c:=2: d:=1:
> tm[1];vm;

$$\frac{1}{3}\sqrt{3}\sqrt{2}$$

$$1+\frac{4}{9}\sqrt{3}\sqrt{2}$$

In the present case, it is fairly clear that $v = 1 + \dfrac{4\sqrt{6}}{9} = 2.0887$ is a maximum. It never hurts to check, though. An easy way to do this is to substitute neighboring values of t:

> evalf(subs(t=tm[1]-0.01,v));evalf(subs(t=tm[1]+0.01,v));
                2.088418159

                2.088416159

Both values are somewhat less than 2.0887, verifying that we have found a maximum. Probably the safest way to verify a maximum, though, is to plot the results.

> plot(v,t=0..1.8, labels=[`time`,`velocity`]);

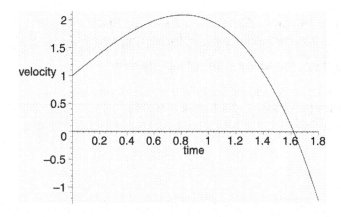

A simpler way to find maxima and minima is to use the exrema function.

> extrema(v,{},t);
$$\{1 - \frac{4}{9}\sqrt{6},\ 1 + \frac{4}{9}\sqrt{6}\}$$

Note that Maple has found both a maximum and a minimum. Further investigation would show that the minimum occurs at the negative time root found above. The brackets {} in the extrema function allow the input of constraints. The problems that we will consider will not have constraints, thus the brackets will be left empty.

Systems of Equations

> restart;
> eqn1:=-P*sin(beta)+Bx-Ax=0;
  eqn2:=Ay+P*cos(beta)-w*a=0;
  eqn3:=P*a*cos(beta)-P*b*sin(beta)+Bx*c-1/2*w*a^2=0;

$$eqn1 := -P\sin(\beta) + Bx - Ax = 0$$

$$eqn2 := Ay + P\cos(\beta) - w\,a = 0$$

$$eqn3 := P\,a\cos(\beta) - P\,b\sin(\beta) + Bx\,c - \frac{1}{2}w\,a^2 = 0$$

Above we have three equations that we are going to solve for three variables using Maple's solve command. Since we have more unknowns than we have equations, Maple will solve the equations symbolically in terms of the unknown parameters. We can ask Maple to solve for any three of the unknowns we wish. Here we will ask Maple to solve for Ax, Ay, and Bx and will obtain three expressions in terms of the remaining parameters P, w, a, b, c and β .

> soln:=solve({eqn1,eqn2,eqn3},{Ax,Ay,Bx});
$$soln := \{Ay = -P\cos(\beta) + w\,a,\ Bx = \frac{1}{2}\frac{-2\,P\,a\cos(\beta) + 2\,P\,b\sin(\beta) + w\,a^2}{c},$$

$$Ax = \frac{1}{2}\frac{-2\,P\sin(\beta)\,c - 2\,P\,a\cos(\beta) + 2\,P\,b\sin(\beta) + w\,a^2}{c}\}$$

It is important to understand how Maple goes about assigning numbers or symbolic expressions to names. Even though the above output has the form Ax = expression, Maple has not yet assigned anything to the name Ax. To see this we type Ax; in the next line and Maple returns simply Ax indicating that nothing is assigned to the name Ax.

> Ax;
$$Ax$$

The assign statement is used to assign the names Ax, Ay and Bx to expressions in soln:

> assign(soln);

The following statements are unnecessary, merely showing explicitly that the assignments have been made.

> Ax; Ay; Bx;

$$\frac{1}{2}\frac{-2\,P\sin(\beta)\,c - 2\,P\,a\,\cos(\beta) + 2\,P\,b\,\sin(\beta) + w\,a^2}{c}$$

$$-P\cos(\beta) + w\,a$$

$$\frac{1}{2}\frac{-2\,P\,a\,\cos(\beta) + 2\,P\,b\,\sin(\beta) + w\,a^2}{c}$$

We are now ready to do a parametric study. As one example, we will assume values for P, w, a, b and c and then plot Ax, Ay, and Bx as a function of β .

> P:=300:w:=50:a:=10:b:=5:c:=20:
> with(plots):
> t1:=textplot([4.2,450,`A x`]):t2:=textplot([2.7,350,`B x`]):t3:=textplot([3.2,870,`A y`]):
> plot1:=plot([Ax,Ay,Bx],beta=0..2*Pi):
> display(plot1,t1,t2,t3);

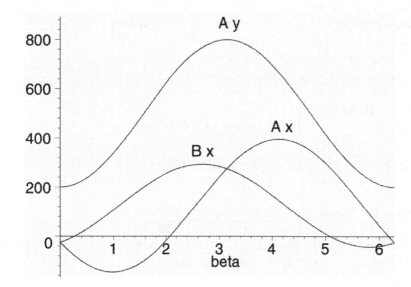

FORCE SYSTEMS

2

This chapter introduces the basic properties of forces and moments in two and three dimensions. Problem 2.1 is a 2D problem that investigates the effects of the orientation of a force upon the resultant of two forces. The problem illustrates how the automatic substitution performed by a computer can simplify what might normally be a rather tedious algebra problem. Problem 2.2 and 2.3 are parametric studies involving 2D moment and equivalent force-couple systems respectively. Problems 2.4 and 2.5 are 3D problems. Problem 2.5 shows how to carry out a cross product with Maple and also involves an interesting design application. Problem 2.6 is another 2D problem that illustrates how to find the maximum value of a moment using *diff* and *solve*.

Following is a summary of the Maple commands used in this chapter.

Table 2.1 Maple Commands used in Chapter 2

Maple Command	Problem
crossprod	2.5
diff	2.6
display	2.1, 2.2, 2.5
evalf	2.6
plot (normal)	2.1, 2.2, 2.3, 2.4, 2.5, 2.6
solve	2.2, 2.6
textplot	2.1, 2.2, 2.5

2.1 Problem 2/20 (2D Rectangular Components)

It is desired to remove the spike from the timber by applying force along its horizontal axis. An obstruction A prevents direct access, so that two forces, one 400 lb and the other **P** are applied by cables as shown. Here we want to investigate the effects of the distance between the spike and the obstruction on the two forces so replace 8" by d in the figure. Determine the magnitude of **P** necessary to insure a resultant **T** directed along the spike. Also find T. Plot P and T as a function of d letting d range between 2 and 12 inches.

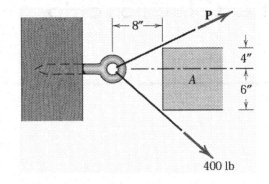

Problem Formulation

The two forces and their resultant are shown on the diagram to the right. The horizontal component of the resultant is T while the vertical component is 0. Thus,

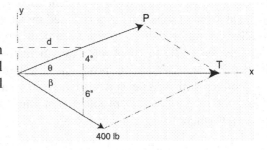

$$T = \Sigma F_x = P\cos\theta + 400\cos\beta$$
$$0 = \Sigma F_y = P\sin\theta - 400\sin\beta$$

These two equations can be easily solved for P and T.

$$P = \frac{400\sin\beta}{\sin\theta} \qquad\qquad T = 400(\sin\beta\cot\theta + \cos\beta)$$

At this point we have P and T as functions of β and θ. From the figure above we can relate β and θ to d as follows.

$$\theta = \tan^{-1}(4/d) \qquad\qquad \beta = \tan^{-1}(6/d)$$

One nice thing about using a computer is that it will not be necessary to substitute these results into those above to get P and T explicitly as functions of d. The computer carries out this substitution automatically. A nice feature of Maple is that it will also make the substitution symbolically providing the explicit dependence of P and T upon d.

Maple Worksheet

> restart;with(plots):

> theta:=arctan(4/d); beta:=arctan(6/d);

$$\theta := \arctan\left(4\,\frac{1}{d}\right)$$

$$\beta := \arctan\left(6\,\frac{1}{d}\right)$$

> T:=400*(sin(beta)*cot(theta)+cos(beta));

$$T := 1000\,\frac{1}{\sqrt{1+\dfrac{36}{d^2}}}$$

> P:=400*sin(beta)/sin(theta);

$$P := 600\,\frac{\sqrt{1+\dfrac{16}{d^2}}}{\sqrt{1+\dfrac{36}{d^2}}}$$

> t1:=textplot([8,840,'T']): t2:=textplot([8,570,'P']):
> p1:=plot([P,T],d=2..12,color=black,labels=[`d (in)`,`Force (lb)`],
labeldirections=[HORIZONTAL,VERTICAL]):
> display(p1,t1,t2);

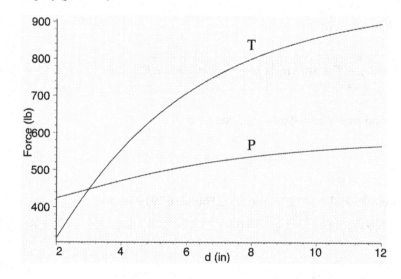

2.2 Problem 2/53 (2D Moment)

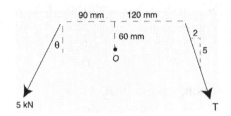

The masthead fitting supports the two forces shown. Here we want to investigate the effects of the orientation of the 5 kN load so replace the angle 30° by θ. Find the magnitude of **T** which will cause no bending of the mast (zero moment) at O. For this T, determine the magnitude of the resultant **R** of the two forces. Plot T and R as a function of θ letting q vary between 0 and 180°.

Problem Formulation

Setting the moment about O to zero yields (note that $\sqrt{2^2 + 5^2} = \sqrt{29}$)

$$M_0 = 5[\cos\theta(90) + \sin\theta(60)] - T\left[\frac{5}{\sqrt{29}}(120) + \frac{2}{\sqrt{29}}(60)\right] = 0$$

Solving, $T = \dfrac{5[\cos\theta(90) + \sin\theta(60)]}{\dfrac{5}{\sqrt{29}}(120) + \dfrac{2}{\sqrt{29}}(60)} = \dfrac{5\sqrt{29}}{24}(3\cos\theta + 2\sin\theta)$

Now we can find the resultant R in terms of T,

$$R_x = \frac{2}{\sqrt{29}}T - 5\sin\theta \;(\rightarrow) \quad R_y = \frac{5}{\sqrt{29}}T + 5\cos\theta \;(\downarrow) \qquad R = \sqrt{R_x^2 + R_y^2}$$

We'll let Maple substitute for T and plot T and R as a function of θ.

Maple Worksheet

```
> restart;with(plots):
> M0:=5*(cos(theta)*90+sin(theta)*60)-T*(5*120/sqrt(29)+2*60/sqrt(29))=0;
```
$$M0 := 450\cos(\theta) + 300\sin(\theta) - \frac{720}{29}T\sqrt{29} = 0$$

```
> T:=solve(M0,T);
```
$$T := \frac{5}{24}(3\cos(\theta) + 2\sin(\theta))\sqrt{29}$$

> Rx:=2/sqrt(29)*T-5*sin(theta);

$$Rx := \frac{5}{4}\cos(\theta) - \frac{25}{6}\sin(\theta)$$

> Ry:=5/sqrt(29)*T+5*cos(theta);

$$Ry := \frac{65}{8}\cos(\theta) + \frac{25}{12}\sin(\theta)$$

> R:=sqrt(Rx^2+Ry^2);

$$R := \sqrt{\left(\frac{5}{4}\cos(\theta) - \frac{25}{6}\sin(\theta)\right)^2 + \left(\frac{65}{8}\cos(\theta) + \frac{25}{12}\sin(\theta)\right)^2}$$

> t1:=textplot([1,4.4,'T']): t2:=textplot([1,7.7,'R']):
> p:=plot([T,R],theta=0..Pi,labels=[`theta (rad)`,`Force(kN)`],
labeldirections=[HORIZONTAL,VERTICAL]):
> display(p,t1,t2);

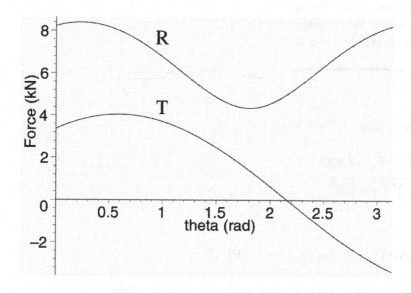

2.3 Problem 2/88 (2D Resultants)

The directions of the two thrust vectors of an experimental aircraft can be independently changed form the conventional forward direction within limits. Here we want to investigate the effects of the orientation of one of the thrusts so change the 15° angle in the figure to θ. For this case determine the equivalent force-couple system at O. Then replace this force-couple system by a single force and specify the point on the x-axis through which the line of action of this resultant passes. Plot (a) the ratio of the magnitude of the resultant to that of the thrust (R/T) and (b) the distance x where the resultant intersects the x-axis as functions of θ for θ between 0 and 30°.

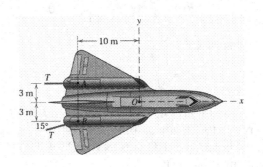

Problem Formulation

For an equivalent force-couple system at O we have,

$$R_x = T + T\cos\theta = T(1+\cos\theta) \quad R_y = T\sin\theta$$

$$R = \sqrt{R_x^2 + R_y^2} = T\sqrt{(1+\cos\theta)^2 + (\sin\theta)^2}$$

$$\frac{R}{T} = \sqrt{(1+\cos\theta)^2 + (\sin\theta)^2}$$

$$M_0 = -T(3) + T\cos\theta(3) - T\sin\theta(3) = T(3\cos\theta - 3 - 10\sin\theta) \ \circlearrowleft$$

Now we want a single resultant force that is equivalent to the force-couple system above. The easiest way to visualize this is to imagine moving the resultant from O to the right along the x-axis. As you do this you generate a counter-clockwise moment about O of $R_y x$ (R_x does not produce a moment about O). The condition for determining x is thus $M_O = R_y x$.

$$x = \frac{M_O}{R_y} = \frac{(3\cos\theta - 3 - 10\sin\theta)}{\sin\theta}$$

Maple Worksheet

> restart;

Let Rp be the ratio R/T

> Rp:=sqrt((1+cos(theta))^2+(sin(theta))^2);

$$Rp := \sqrt{(1 + \cos(\theta))^2 + \sin(\theta)^2}$$

> plot(Rp,theta=0..30*Pi/180,y=1.8..2, labels=[`theta (rads)`, `R/T`]);

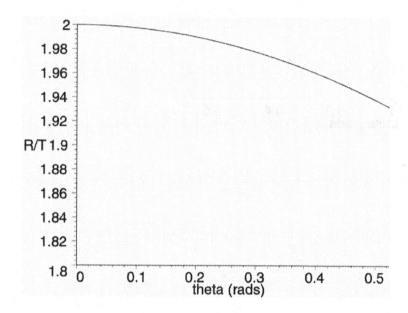

> x:=(3*cos(theta)-3-10*sin(theta))/sin(theta);

$$x := \frac{3 \cos(\theta) - 3 - 10 \sin(\theta)}{\sin(\theta)}$$

> plot(x, theta=0..30*Pi/180,y=-10..-11, labels=[`theta (rads)`, `x (m)`]);

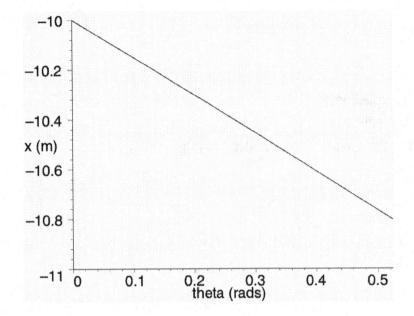

2.4 Problem 2/105 (3D Rect. Components)

The rigid pole and cross-arm assembly is supported by the three cables shown. A turnbuckle at D is tightened until it induces a tension T in CD of 1.2 kN. Let the distance OD be d instead of 3 m and determine the magnitude T_{GF} of the projection of **T** onto line GF. Plot T_{GF} as a function of d letting d vary between 0 and 20 meters.

Problem Formulation

First we write **T** as a Cartesian vector and then find a unit vector along line GF.

$$\mathbf{T} = T\frac{1.5\mathbf{i} + d\mathbf{j} - 4.5\mathbf{k}}{\sqrt{(1.5)^2 + d^2 + (4.5)^2}} = (1.2)\frac{1.5\mathbf{i} + d\mathbf{j} - 4.5\mathbf{k}}{\sqrt{22.5 + d^2}}$$

$$\mathbf{n}_{GF} = \frac{2\mathbf{i} - 3\mathbf{k}}{\sqrt{(2)^2 + (3)^2}} = \frac{2\mathbf{i} - 3\mathbf{k}}{\sqrt{13}}$$

The projection T_{GF} is now found by taking the dot product of the above two vectors.

$$T_{GF} = \mathbf{T} \cdot \mathbf{n}_{GF} = \frac{1.2(1.5)(2) + 1.2d(0) + 1.2(-4.5)(-3)}{\sqrt{13}\sqrt{22.5 + d^2}} = \frac{19.8/\sqrt{13}}{\sqrt{22.5 + d^2}}$$

Maple Worksheet

```
> restart;
> Tgf:=19.8/sqrt(13)/sqrt(22.5+d^2);
```

$$Tgf := 1.523076923\frac{\sqrt{13}}{\sqrt{22.5 + d^2}}$$

```
> plot(Tgf,d=0..20,y=0..1.2, labels=[`d (m)`,`Tgf (kN)`],
labeldirections=[HORIZONTAL,VERTICAL]);
```

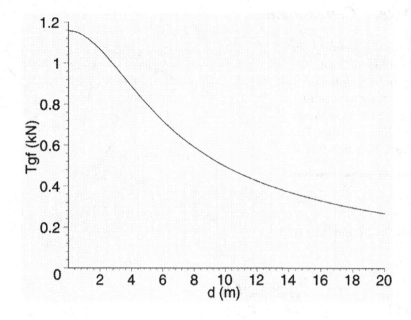

2.5 Sample Problem 2/13 (3D Moment)

A tension **T** of magnitude 10 kN is applied to the cable attached to the top A of the rigid mast and secured to the ground at B. Let the coordinates of point B be $(x_B, 0, z_B)$ instead of $(12, 0, 9)$ as in the figure. Find a general expression for $\mathbf{M_O}$ (the moment of **T** about the base O) as a function of x_B and z_B. (a) Plot the magnitude of $\mathbf{M_O}$ (M_O) and its components about the x and z axes (M_x and M_z) as a function of x_B for $z_B = 9$ m. For this case, determine the suitable range for x_B if M_O is not to exceed 100 kN•m. (b) Find a simple way to explain to a construction worker (who has not taken this course) all safe locations for B if M_O cannot exceed 100 kN•m.

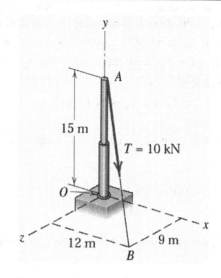

Problem Formulation

First, we express T as a Cartesian vector and then find the moment by taking a cross product. In the worksheet below, Maple will be used to evaluate the cross product.

$$\mathbf{T} = T\mathbf{n}_{AB} = 10\left[\frac{x_B\mathbf{i} - 15\mathbf{j} + z_B\mathbf{k}}{\sqrt{x_B^2 + 15^2 + z_B^2}}\right]$$

$$\mathbf{M_O} = \mathbf{r}_{OA} \times \mathbf{T} = 15\mathbf{j} \times 10\left[\frac{x_B\mathbf{i} - 15\mathbf{j} + z_B\mathbf{k}}{\sqrt{x_B^2 + 15^2 + z_B^2}}\right] = 150\left[\frac{z_B\mathbf{i} - x_B\mathbf{k}}{\sqrt{x_B^2 + 15^2 + z_B^2}}\right]$$

(a) M_x and M_z are the scalar components of $\mathbf{M_O}$ in the x and z-directions respectively.

$$M_x = \frac{150z_B}{\sqrt{x_B^2 + 15^2 + z_B^2}} \qquad M_z = \frac{-150x_B}{\sqrt{x_B^2 + 15^2 + z_B^2}}$$

$$M_O = \sqrt{M_x^2 + M_z^2} = 150\sqrt{\frac{x_B^2 + z_B^2}{x_B^2 + 15^2 + z_B^2}}$$

After substituting $z_B = 9$, M_O, M_x, and M_z can be plotted as a function of x_B. The result can be found in the Maple worksheet that follows. To find the acceptable range of x_B given $M_O \leq 100$ kN•m, substitute $M_O = 100$ and $z_B = 9$ into the equation for M_O and then solve for x_B. The result is $x_B = \pm 3\sqrt{11}$ m. Thus, the acceptable range is

$$-3\sqrt{11} \leq x_B \leq 3\sqrt{11} \text{ m}$$

(b) Substituting $M_O = 100$ into the equation above gives,

$$100 = 150\sqrt{\frac{x_B^2 + z_B^2}{x_B^2 + 15^2 + z_B^2}} \ .$$

Squaring both sides of this equation and rearranging terms yields

$$x_B^2 + z_B^2 = 180$$

which is a circle of radius $\sqrt{180} = 13.42$ m (44 ft). Thus, the construction worker should be instructed to keep B within 44 feet of the mast.

Maple Worksheet

> restart; with(linalg): with(plots):
> rab:=sqrt(xb^2+15^2+zb^2);
$$rab := \sqrt{xb^2 + 225 + zb^2}$$

> Tab:=vector([10*xb/rab,-15*10/rab,zb*10/rab]); # defines T as a vector.
$$Tab := \left[10\frac{xb}{\sqrt{xb^2 + 225 + zb^2}}, -150\frac{1}{\sqrt{xb^2 + 225 + zb^2}}, 10\frac{zb}{\sqrt{xb^2 + 225 + zb^2}} \right]$$

> M0:=crossprod([0,15,0],Tab); # evaluates the cross product.
$$M0 := \left[150\frac{zb}{\sqrt{xb^2 + 225 + zb^2}}, 0, -150\frac{xb}{\sqrt{xb^2 + 225 + zb^2}} \right]$$

> Mx:=M0[1]; Mz:=M0[3];
$$Mx := 150\frac{zb}{\sqrt{xb^2 + 225 + zb^2}}$$

$$Mz := -150\frac{xb}{\sqrt{xb^2 + 225 + zb^2}}$$

> M0m:=sqrt(Mx^2+Mz^2);
$$M0m := 150\sqrt{\frac{zb^2}{xb^2 + 225 + zb^2} + \frac{xb^2}{xb^2 + 225 + zb^2}}$$

> zb:=9;
$$zb := 9$$

> t1:=textplot([10,110,"Mo"]): t2:=textplot([10,75,"Mx"]):
  t3:=textplot([10,-60,"Mz"]):
> p:=plot([M0m,Mx,Mz], xb=-15..15,labels=[`xB (m)`,``],title="M (kN m)"):
> display(p,t1,t2,t3);

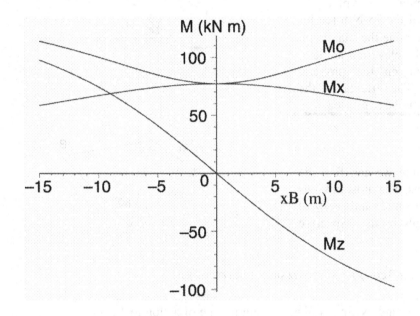

> solve(M0m=100,xb);
$$3\sqrt{11},\ -3\sqrt{11}$$

2.6 Problem 2/181 (2D Moment)

A flagpole with attached light triangular frame is shown here for an arbitrary position during its raising. The 75-N tension in the erecting cable remains constant. Determine and plot the moment about the pivot O of the 75-N force for the range $0 \le \theta \le 90°$. Determine the maximum value of this moment and the elevation angle at which it occurs; comment on the physical significance of the latter. The effects of the diameter of the drum at D may be neglected.

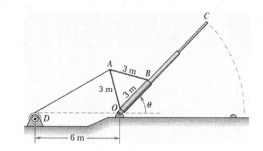

Problem Formulation

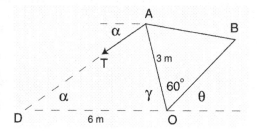

From the diagram to the right we can find the moment about O in two ways. The most obvious is to resolve T into horizontal and vertical components and then multiply by the respective moment arms. The result is,

$$M_O = T\cos\alpha(3\sin\gamma) + T\sin\alpha(3\cos\gamma) = 3T(\cos\alpha\sin\gamma + \sin\alpha\cos\gamma)$$

A simpler expression can be found by first sliding T along its line of action to D. Once this has been done we see that only the vertical component ($T\sin\alpha$) produces a moment about O with moment arm of 6 m. Thus,

$$M_O = 6T\sin\alpha$$

Here we will use the second, simpler, expression. As an exercise you may want to verify that the two expressions yield identical results.

Now we have to do a little geometry to relate α to θ. From the diagram we have,

$$\gamma + 60° + \theta = 180° \qquad \gamma = 120° - \theta \qquad \gamma = \frac{2\pi}{3} - \theta \text{ (radians)}$$

$$AD = \sqrt{6^2 + 3^2 - 2(6)(3)\cos\gamma} \qquad\qquad \text{(law of cosines)}$$

$$\frac{\sin\alpha}{3} = \frac{\sin\gamma}{AD} \qquad\qquad \alpha = \sin^{-1}\left(\frac{3\sin\gamma}{AD}\right) \qquad\qquad \text{(law of sines)}$$

This problem illustrates very well one of the advantages of a computer solution. At this point you should be sure that you understand that we now have the moment as a function of θ. The reason is that we have M_O as a function of α, α as a function of γ and γ as a function of θ. It is not necessary to actually make the substitutions ourselves, the computer will do that for us. In addition to this, Maple, unlike many other programs, will also make the substitution symbolically. For example, from the worksheet below, we have the following symbolic result for M_O,

$$M_O = \frac{450\sin\left(\dfrac{\pi}{3}+\theta\right)}{\sqrt{5+4\cos\left(\dfrac{\pi}{3}+\theta\right)}}$$

The maximum moment and the elevation at which it occurs are determined in the worksheet below. The angle θ is determined by solving the equation $dM_O/d\theta = 0$. Substitution of this result into M_O gives the maximum moment. The results are $M_{Omax}= 225$ N-m at $\theta = \pi/3 = 60°$.

It turns out that we could have anticipated this result with some simple geometry. First, we know that the maximum moment will occur when **T** is perpendicular to OA. This yields the right triangle shown to the right. From this diagram we see that $\cos\gamma = 3/6$ so that $\gamma = 60°$. From our results above, $\theta = 60°$ when $\gamma = 60°$. Also, for this orientation, $M_O = 3(75) = 225$ N-m.

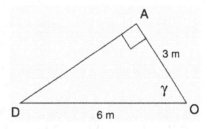

Maple Worksheet

> restart; unprotect(gamma);
> gamma:=120*Pi/180-theta;

$$\gamma := \frac{2}{3}\pi - \theta$$

> AD:=sqrt(6^2+3^2-2*6*3*cos(gamma));

$$AD := 3\sqrt{5+4\cos\left(\frac{1}{3}\pi+\theta\right)}$$

> alpha:=arcsin(3*sin(gamma)/AD);

$$\alpha := \arcsin\left(\frac{\sin\left(\frac{1}{3}\pi + \theta\right)}{\sqrt{5 + 4\cos\left(\frac{1}{3}\pi + \theta\right)}}\right)$$

> M0:=6*75*sin(alpha);

$$M0 := 450\,\frac{\sin\left(\frac{1}{3}\pi + \theta\right)}{\sqrt{5 + 4\cos\left(\frac{1}{3}\pi + \theta\right)}}$$

> plot(M0,theta=0..Pi/2, labels=["theta (rad)"," "],title="Moment of T about O (N-m)");

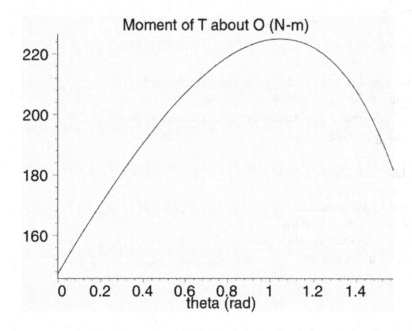

To find the maximum moment we first differentiate Mo with respect to theta and set the result equal to zero. Solving this equation gives the angle theta at which the maximum occurs. This value is then substituted back into Mo to get the maximum.

> dM0:=diff(M0,theta);

$$dM0 := 450\,\frac{\cos\left(\frac{1}{3}\pi + \theta\right)}{\sqrt{5 + 4\cos\left(\frac{1}{3}\pi + \theta\right)}} + \frac{900\,\sin\left(\frac{1}{3}\pi + \theta\right)^2}{\left(5 + 4\cos\left(\frac{1}{3}\pi + \theta\right)\right)^{(3/2)}}$$

> solve(dM0=0,theta);

$$\frac{\pi}{3}, \frac{2\pi}{3} - \arccos(2)$$

Of the two solutions Maple has found, only the first (Pi/3 = 60) is real.

> evalf(subs(theta=1/3*Pi,M0));

      225.0000000

EQUILIBRIUM

3

This chapter considers equilibrium of two and three dimensional structures and is the foundation for the study of engineering Statics. Problems 3.1 through 3.3 are two dimensional equilibrium problems while problem 3.4 involves equilibrium in three dimensions. Since equilibrium problems usually involve solving the equilibrium equations for unknown forces it is not surprising that all but one of the problems in this chapter use the Maple *solve* command. In particular, problem 3.4 solves three equations symbolically for three unknowns. Problem 3.3 illustrates how to find the maximum and minimum values of a force by using Maple's *diff* and *solve* commands. The same results are then found using *maximize* and *minimize*. In problem 3.4, crossprod is used to symbolically evaluate a cross product. A few explanatory notes on vector algebra are included in this problem since this topic was not covered in chapter 1.

Following is a summary of the Maple commands used in this chapter.

Table 3.1 Maple Commands used in Chapter 3

Maple Command	Problem
crossprod	3.4
diff	3.3
display	3.2, 3.3, 3.4
maximize	3.3
minimize	3.3
plot (normal)	3.1, 3.2, 3.3, 3.4
simplify	3.3
solve	3.1, 3.3, 3.4
subs	3.3
textplot	3.2, 3.3,3.4

3.1 Problem 3/26 (2D Equilibrium)

The indicated location of the center of gravity of the 3600-lb pickup truck is for the unladen condition. A load W_L whose center of gravity is x inches behind the rear axle is added to the truck. Find the relationship between W_L and x if the normal forces under the front and rear wheels are to be equal. For this case, plot W_L as a function of x for x ranging between 0 and 50 inches.

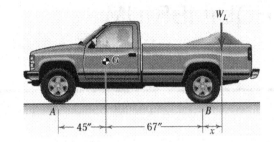

Problem Formulation

The free-body diagram for the truck is shown to the right. Normally, the two normal forces under the wheels would not be identical, of course. Here we want the relationship between the weight W_L and its location (x) which results in these two forces being equal. This relationship is found from the equilibrium equations.

$$\Sigma M_B = 0 = 3600(67) - N(112) - W_L x = 0$$

$$\Sigma F_y = 0 = N + N - 3600 - W_L = 0$$

The second equation can be solved for N and then substituted into the first equation to yield the required relation between W_L and x. This relation can then be solved for W_L.

Maple Worksheet

```
> restart;
> sMb:=3600*67-N*112-WL*x=0;
            sMb := 241200 - 112 N - WL x = 0

> sFy:=2*N-3600-WL=0;
            sFy := 2 N - 3600 - WL = 0

> N:=solve(sFy,N);
            N := 1800 + 1/2 WL
```

At this point Maple will automatically substitute N back into the first equation as verified below.

> sMb;
$$39600 - 56\,WL - WLx = 0$$

Now we solve this equation for WL.

> WL:=solve(sMb,WL);
$$WL := 39600\,\frac{1}{56 + x}$$

> plot(WL, x=0..50, y=300..800, labels=["x (in)",""],title="Load weight (lb)");

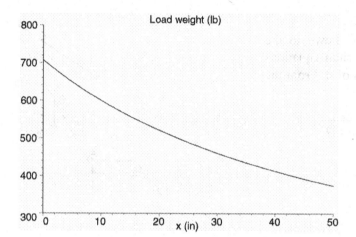

3.2 Problem 3/37 (2D Equilibrium)

The uniform 18-kg bar OA is held in the position shown by the smooth pin at O and the cable AB. Determine the tension T in the cable and the magnitude of the external pin reaction at O in terms of the angle θ. Plot the forces as a function of θ for $15^{0} \leq \theta \leq 90^{0}$.

Problem Formulation

The free-body diagram for the bar is shown to the right. Before proceeding to the equilibrium equations we must first find the angle α in terms of θ. From the law of cosines and law of sines,

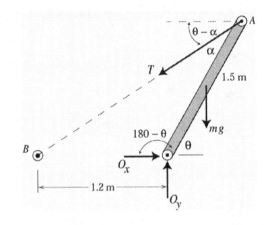

$$AB = \sqrt{1.2^{2} + 1.5^{2} - 2(1.2)(1.5)\cos(\pi - \theta)}$$

$$\frac{\sin \alpha}{1.2} = \frac{\sin(\pi - \theta)}{AB}$$

Observing that $\cos(\pi - \theta) = -\cos\theta$ and $\sin(\pi - \theta) = \sin\theta$,

$$AB = \sqrt{3.69 + 3.60\cos\theta}$$

$$\alpha = \sin^{-1}\left(\frac{1.2\sin\theta}{AB}\right)$$

The equilibrium equations can now be written,

$$\circlearrowleft \Sigma M_{O} = 0 = mg\left(\frac{1.5}{2}\cos\theta\right) - T\sin\alpha(1.5)$$

$$\Sigma F_{x} = 0 = O_{x} - T\cos(\theta - \alpha)$$

$$\Sigma F_{y} = 0 = O_{y} - mg - T\sin(\theta - \alpha)$$

The three equations above can be solved for the forces,

$$T = \frac{mg}{2} \frac{\cos\theta}{\sin\alpha}$$

$$O_x = T\cos(\theta - \alpha) \qquad\qquad O_y = mg + T\sin(\theta - \alpha)$$

$$O = \sqrt{O_x^2 + O_y^2}$$

Maple Worksheet

```
> restart; with(plots):
> AB:=sqrt(1.2^2+1.5^2-2*1.2*1.5*cos(Pi-theta));
```
$$AB := \sqrt{3.69 + 3.60\,\cos(\theta)}$$

```
> alpha:=arcsin(1.2/AB*sin(Pi-theta));
```
$$\alpha := \arcsin\left(\frac{1.2\,\sin(\theta)}{\sqrt{3.69 + 3.60\,\cos(\theta)}}\right)$$

```
> T:=W/2*cos(theta)/sin(alpha);
```
$$T := \frac{0.4166666666\ \ W\,\cos(\theta)\,\sqrt{3.69 + 3.60\,\cos(\theta)}}{\sin(\theta)}$$

```
> Ox:=T*cos(theta-alpha):
> Oy:=W+T*sin(theta-alpha):
> Om:=sqrt(Ox^2+Oy^2):
> W:=18*9.81;
```
$$W := 176.58$$

```
> p1:=plot([T,Om],theta=15*Pi/180..Pi/2,color=black,
      labels=["theta (radians)",""],title="Force (N)"):
> t1:=textplot([1,290,"O"]):t2:=textplot([1,150,"T"]):
> display(p1,t1,t2);
```

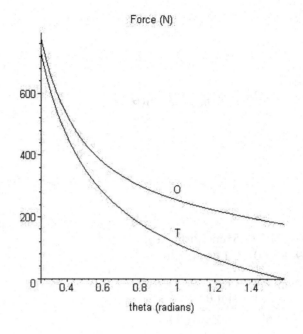

3.3 Sample Problem 3/4 (2D Equilibrium)

Let the 10 kN load be located a distance c (meters) to the left of B. Find the magnitude T of the tension in the supporting cable and the magnitude of the force on the pin at A in terms of c. (a) Plot T and A as a function of c letting c range between 0 and 5 m. (b) Find the minimum and maximum values for T and A when c varies between 0 and 5 m. The beam AB is a standard 0.5-m I-beam with a mass of 95 kg per meter of length.

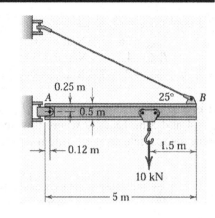

Problem Formulation

The free-body diagram is shown in the figure to the right. The weight of the beam is $(95)(5)(9.81) = 4660$ N or 4.66 kN. The weight acts at the center of the beam. The equilibrium equations are now written from the free-body diagram.

$$[\Sigma M_A = 0] \qquad (T\cos 25°)0.25 + (T\sin 25°)(5 - 0.12)$$
$$- 10(5 - c - 0.12) - 4.66(2.5 - 0.12) = 0$$

from which $\ T = 26.15 - 4.367c$

$$[\Sigma F_x = 0] \quad A_x - T\cos 25° = 0$$
$$A_x = (26.15 - 4.367c)\cos 25° = 23.70 - 3.958c$$

$$[\Sigma F_y = 0] \quad A_y + T\sin 25° - 4.66 - 10 = 0$$
$$A_y = 5.66 - (26.15 - 4.367c)\sin 25° = 3.61 + 1.846c$$

$$A = \sqrt{A_x^2 + A_y^2} = \sqrt{(23.7 - 3.958c)^2 + (3.61 + 1.846c)^2}$$
$$A = \sqrt{574.7 - 174.3c + 19.07c^2}$$

(a) The plot of T and A as a function of c can be found in the worksheet below.

(b) The locations c for the maximum values of T and A and the minimum value for T are clear from the plot below. Substituting these values of c into the equations above yields,

$T_{max} = 26.15$ kN at $c = 0$
$T_{min} = 4.31$ kN at $c = 5$ m
$A_{max} = 23.97$ kN at $c = 0$

It is also clear from the plot that A goes through a minimum somewhere between $c = 4$ and 5 m. Exactly where this occurs can be determined by differentiating A with respect to c and equating the result to zero.

$$\frac{dA}{dc} = \frac{19.07c - 87.15}{\sqrt{574.7 - 174.3c + 19.07c^2}} = 0$$

from which $c = 4.57$ m. Substituting this value into A yields,

$A_{min} = 13.28$ kN at $c = 4.57$ m

As we will see in the worksheet below, Maple can be used to obtain many of the detailed aspects of the solution outlined above.

Maple Worksheet

```
> restart; with(plots):
> Digits:=4:
> theta:=evalf(25*Pi/180):
> eqn1:=T*cos(theta)*0.25+T*sin(theta)*(5-0.12)-10*(5-c-0.12)-4.66*(2.5-0.12);
        eqn1 := 2.290 T − 59.89 + 10 c

> T:=solve(eqn1,T);
        T := 26.15 − 4.367 c

> Ax:=T*cos(theta);
        Ax := 23.70 − 3.958 c

> Ay:=14.66-T*sin(theta);
        Ay := 3.61 + 1.846 c

> A:=simplify(sqrt(Ax^2+Ay^2));
        A := .1000√(57470. − 17430. c + 1907. c²)

> p:=plot([T,A],c=0..5, y=0..30, labels=[`c (m)`,`Force (kN)`], labeldirections =
[HORIZONTAL,VERTICAL]):
> t1:=textplot([4,15,"A"]): t2:=textplot([4,10.4,"T"]):
> display(p,t1,t2);
```

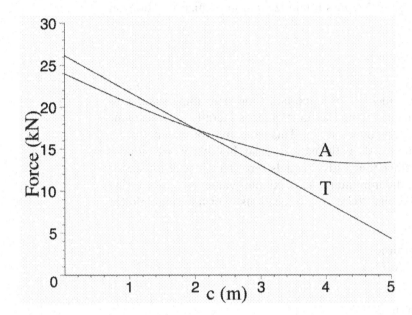

The locations for the maximum values of T and A and the minimum value for T are clear from the plot. We can obtain the associated maximum and minimum values by substitution.

> Tmax:=subs(c=0,T); Tmin:=subs(c=5,T); Amax:=subs(c=0,A);

$$Tmax := 26.15$$

$$Tmin := 4.31$$

$$Amax := 23.97$$

To find the location of Amin we first differentiate A with respect to c and set the result equal to zero. The value of c thus determined is then substituted back into A to yielding Amin.

> diff(A,c);

$$.05000 \frac{-17430.+3814.c}{\sqrt{57470.-17430.c+1907.c^2}}$$

> solve(%=0,c); # % is a shortcut for the last expression evaluated in Maple
4.570

> Amin:=subs(c=%,A);

$$Amin := 13.28$$

The above represents the usual method for finding a minimum of a function. It is also possible to take advantage of Maple's minimize and maximize functions as illustrated below.

> minimize(A,c=0..5,location=true);

$$13.28, \{[\{c = 4.570\}, 13.28]\}$$

This is, of course, very convenient. It is important though to understand how these functions work before using them. The function does literally what its name implies. It finds the smallest value over a range. This value may or may not occur at a relative max or min in the usual, mathematical sense. In other words, it does not find locations where the slope (derivative) equals zero. This is well illustrated by using the function to find the minimum and maximum values of T. Referring back to the graph you will see that there are no locations over the range plotted where dT/dc = 0.

> maximize(T,c=0..5,location=true);

$$26.15, \{[\{c = 0\}, 26.15]\}$$

> minimize(T,c=0..5,location=true);

$$4.31, \{[\{c = 5\}, 4.31]\}$$

3.4 Sample Problem 3/7 (3D Equilibrium)

The welded tubular frame is secured to the horizontal x-y plane by a ball and socket joint at A and receives support from the loose-fitting ring at B. Under the action of the 2-kN load, the cable CD prevents rotation about a line from A to B, and the frame is stable in the position shown. Let the distance from the z-axis to the ring at B be c meters so that the coordinates of B are $(0, c, 6)$. Find expressions for the tension T in the cable and the magnitudes of the forces at A and B in terms of c. Plot T, A, and B as a function of c for $0 \leq c \leq 6$ m. Explain why A and B go to infinity as c approaches 0. You may neglect the weight of the frame.

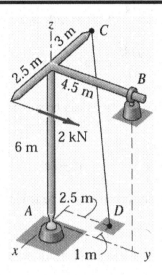

Problem Formulation

In the solution to the sample problem in your text, moments were summed about the AB axis in order to obtain one equation with one unknown, the tension T. When using computer software capable of symbolic algebra such considerations are far less important. In fact, it is often advisable to take the most straightforward approach in setting up the problem and then use the computer to work out the algebra.

Summing moments about A in the free-body diagram shown to the right,

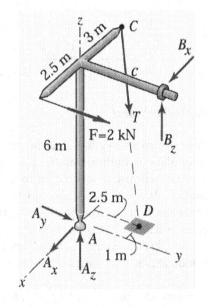

$$[\Sigma M_A = 0] \qquad r_{AD} \times T + r_{AE} \times F + r_{AB} \times B = 0$$

where $T = \dfrac{T(2i + 2.5j - 6k)}{\sqrt{(2)^2 + (2.5)^2 + (-6)^2}} = \dfrac{2T}{\sqrt{185}}(2i + 2.5j - 6k)$

$$F = 2j \ (kN) \qquad B = B_x i + B_z k$$
$$r_{AD} = -i + 2.5j \quad r_{AE} = 2.5i + 6k \quad r_{AB} = cj + 6k$$

Completion of the vector operations above yields a vector equation whose x, y, and z components give three scalar equations for the summation of moments about the x, y, and z axes respectively,

$$\Sigma M_x = -\frac{6\sqrt{185}}{37}T - 12 + cB_z = 0 \qquad \Sigma M_y = -\frac{12\sqrt{185}}{185}T + 6B_x = 0$$

$$\Sigma M_z = -\frac{3\sqrt{185}}{37}T + 5 - cB_x = 0$$

These three equations can be solved simultaneously to give

$$T = \frac{5\sqrt{185}}{15 + 2c} \qquad B_x = \frac{10}{15 + 2c} \qquad B_z = \frac{6(55 + 4c)}{c(15 + 2c)}$$

The remaining unknowns can be found by summing forces,

$$\Sigma F_x = T_x + B_x + A_x = 0; \; \Sigma F_y = T_y + 2 + A_y = 0;$$
$$\Sigma F_z = T_z + B_z + A_z = 0$$

Substituting the components of $\mathbf{T}$ (T_x, T_y, T_z) and $\mathbf{B}$ (B_x, B_z) into the above yields,

$$A_x = -\frac{30}{15 + 2c} \qquad A_y = -\frac{55 + 4c}{15 + 2c} \qquad A_z = \frac{6(6c - 55)}{c(15 + 2c)}$$

Finally, $A = \sqrt{A_x^2 + A_y^2 + A_z^2}$ and $B = \sqrt{B_x^2 + B_z^2}$

Maple Worksheet

There are two ways to solve this problem with Maple. One is to go through the manipulations outlined above by hand and then use Maple to plot the results. Here we will use a second approach that makes better use of Maple's symbolic abilities. We will let Maple do the vector algebra (primarily cross products), which result in the equations summarized above. We will then use Maple to solve those equations. This results in a rather lengthy worksheet but avoids some tedious algebra. Since vector algebra was not covered in Chapter 1 we will also include explanatory notes along the way.

> restart; with(linalg): with(plots): # we need to call Maple's linalg library.

First we set up the three position vectors needed in the cross products.

> rad:=vector([-1,5/2,0]); rae:=vector([5/2,0,6]); rab:=vector([0,c,6]);

$$rad := \left[-1, \frac{5}{2}, 0 \right]$$

$$rae := \left[\frac{5}{2}, 0, 6 \right]$$

$rab := [0, c, 6]$

Now we want to express the tension as a Cartesian vector. We need to be careful to distinguish between the vector and its magnitude. Here we will define Tcd as the vector with magnitude T.

> rcdm:=sqrt((2)^2+(5/2)^2+6^2); # finds the magnitude of the position vector rcd.
$$rcdm := \frac{1}{2}\sqrt{185}$$

> Tcd:=vector([T*2/rcdm,5/2*T/rcdm,-6*T/rcdm]); # defines the tension as a vector.
$$Tcd := \left[\frac{4}{185}T\sqrt{185}, \frac{1}{37}T\sqrt{185}, -\frac{12}{185}T\sqrt{185}\right]$$

Now we define the force F and the reactions at A and B as vectors

> F:=vector([0,2,0]); B:=vector([Bx,0,Bz]); A:=vector([Ax,Ay,Az]);
$$F := [0, 2, 0]$$

$$B := [Bx, 0, Bz]$$

$$A := [Ax, Ay, Az]$$

> crossprod(rad,Tcd)+crossprod(rae,F)+crossprod(rab,B);
$$\left[-\frac{6}{37}T\sqrt{185}, -\frac{12}{185}T\sqrt{185}, -\frac{3}{37}T\sqrt{185}\right] + [-12, 0, 5] + [c\,Bz, 6\,Bx, -c\,Bx]$$

The expression above evaluates the cross products for the summation of moments about A. This is a vector equation where the first, second and third terms in each bracketed expression correspond to our x, y and axes. From this we can get three scalar equations for the summation of moments about the x, y and z axes.

> sumMx:=-6/37*T*sqrt(185) - 12 + c*Bz=0;
$$sumMx := -\frac{6}{37}T\sqrt{185} - 12 + c\,Bz = 0$$

> sumMy:= -12/185*T*sqrt(185)+6*Bx=0;
$$sumMy := -\frac{12}{185}T\sqrt{185} + 6\,Bx = 0$$

> sumMz:=-3/37*T*sqrt(185)+5-c*Bx=0;
$$sumMz := -\frac{3}{37}T\sqrt{185} + 5 - c\,Bx = 0$$

Now we solve these three equations for the three unknowns Bx, Bz and T.

> soln1:=solve({sumMx,sumMy,sumMz},{Bx,Bz,T});

$$soln1 := \{\, T = 5\,\frac{\sqrt{185}}{15+2\,c},\, Bx = 10\,\frac{1}{15+2\,c},\, Bz = 6\,\frac{55+4\,c}{c\,(15+2\,c)}\,\}$$

> assign(soln1);

At this point we already have all forces expressed as vectors. We can thus write three scalar equations for the summation of forces in the x, y, and z directions by taking the first, second, and third terms in each vector. Note how Maple automatically substitutes the results previously solved for.

> sumFx:=Tcd[1]+F[1]+B[1]+A[1]=0;

$$sumFx := 30\,\frac{1}{15+2\,c} + Ax = 0$$

> sumFy:=Tcd[2]+F[2]+B[2]+A[2]=0;

$$sumFy := 25\,\frac{1}{15+2\,c} + 2 + Ay = 0$$

> sumFz:=Tcd[3]+F[3]+B[3]+A[3]=0;

$$sumFz := -60\,\frac{1}{15+2\,c} + \frac{6\,(55+4\,c)}{c\,(15+2\,c)} + Az = 0$$

> soln2:=solve({sumFx,sumFy,sumFz},{Ax,Ay,Az});

$$soln2 := \{\, Az = 6\,\frac{6\,c-55}{c\,(15+2\,c)},\, Ay = -\frac{55+4\,c}{15+2\,c},\, Ax = -30\,\frac{1}{15+2\,c}\,\}$$

> assign(soln2);
> Am:=simplify(sqrt(Ax^2+Ay^2+Az^2)); # finds the magnitude of the force at A.

$$Am := \sqrt{\frac{5221\,c^2 + 440\,c^3 + 16\,c^4 - 23760\,c + 108900}{c^2\,(15+2\,c)^2}}$$

> Bm:=simplify(sqrt(Bx^2+Bz^2)); # finds the magnitude of the force at B.

$$Bm := 2\,\sqrt{\frac{169\,c^2 + 27225 + 3960\,c}{c^2\,(15+2\,c)^2}}$$

Now we are finally ready to make our plot.

> p1:=plot([Am,Bm,T], c=0..10,y=0..10, labels=[`c (m)`,`Force (kN)`],
labeldirections=[HORIZONTAL,VERTICAL]):
> t1:=textplot([2.5 ,10 ,"B"]): t2:=textplot([1.3 ,10 ,"A"]):
  t3:=textplot([1.3 ,4.5 ,"T"]):
> display(p1,t1,t2,t3);

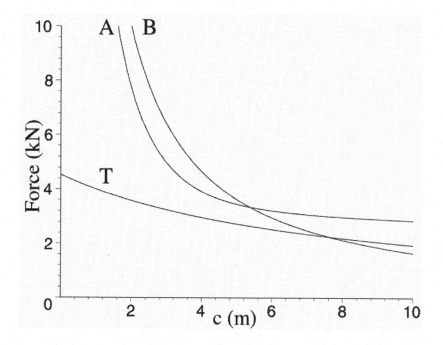

Note that the ring can no longer support the frame when it is located exactly on the z-axis. It is for this reason that the forces A and B go to infinity as c approaches 0.

STRUCTURES

4

This chapter concerns the determination of internal forces in a structure. A structure is an assembly of connected members designed to support or transfer forces. Thus, the internal forces that are generally of interest are the forces of action and reaction between the connected members of the structure. The types of structures considered here are generally classified as trusses, frames or machines.

Problems 4.1 and 4.2 use the methods of joints and sections respectively to analyze a two dimensional truss. Problem 4.1 solves two equations simultaneously using *solve*. Problem 4.3 uses the method of joints on a space truss and utilizes Maple's *solve* command to solve three equations symbolically. Problems 4.4 and 4.5 cover frames and machines.

Following is a summary of the Maple commands used in this chapter.

Table 4.1 Maple Commands used in Chapter 4

Maple Command	Problem
display	4.1, 4.2, 4.3, 4.5
plot (normal)	4.1, 4.2, 4.3, 4.4, 4.5
solve	4.1, 4.3
textplot	4.1, 4.2, 4.3, 4.4, 4.5

4.1 Problem 4/14 (Trusses. Method of Joints)

The truss is composed of equilateral triangles of sides a and is loaded and supported as shown. Determine the reaction force on the roller at E and the forces in members FE, and DE as a function of θ. Plot the non-dimensional loads E/L, FE/L, and DE/L for θ between 0 and 45 degrees.

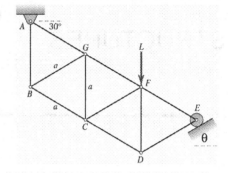

Problem Formulation

First we determine the reaction force at E from a free-body diagram for the entire truss (shown to the right).

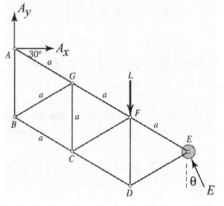

$$\Sigma M_A = -L(2a\cos(30)) + E\cos\theta(3a\cos(30))$$

$$-E\sin\theta(3a\sin(30)) = 0$$

$$E = \frac{2}{3}\frac{L\cos 30}{\cos\theta\cos 30 - \sin\theta\sin 30} = \frac{2}{3}\frac{L\sqrt{3}}{\sqrt{3}\cos\theta - \sin\theta}$$

Note that the non-dimensional force E/L can be easily found by dividing both sides of the equation by L.

$$\frac{E}{L} = \frac{2}{3}\frac{\sqrt{3}}{\sqrt{3}\cos\theta - \sin\theta}$$

Note also that this same equation could be obtained just as easily by setting $L = 1$. Thus one interpretation of a non-dimensional load such as E/L is the force per unit load L.

To obtain the required forces FE and DE we now consider a free-body diagram for joint E. Note that each member is assumed to be in tension. Thus, positive answers will imply tension and negative answers compression.

Joint E

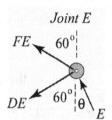

Joint E

$$\Sigma F_x = 0 = -FE\sin 60 - DE\sin 60 - E\sin\theta$$

$$\Sigma F_y = 0 = FE\cos 60 - DE\cos 60 + E\cos\theta$$

Note that after E is substituted that the above are two equations in two unknowns for FE and DE. We will let Maple carry out the solution to these equations in the worksheet below. The result is

$$\frac{FE}{L} = \frac{\tan\theta + \sqrt{3}}{\tan\theta - \sqrt{3}} \qquad\qquad \frac{DE}{L} = \frac{2}{3}$$

Of course, the fact that the force in member DE is 2L/3 independent of θ is probably surprising.

Maple Worksheet

> restart; with(plots):
> SumMA:=-L*2*a*cos(30*Pi/180)+E*cos(theta)*3*a*cos(30*Pi/180)-E*sin(theta)*3*a*sin(30*Pi/180);

$$SumMA := -L\,a\,\sqrt{3} + \frac{3}{2}E\cos(\theta)\,a\,\sqrt{3} - \frac{3}{2}E\sin(\theta)\,a$$

> L:=1: # set L = 1 to find the non-dimensional load E/L
> E:=solve(SumMA=0,E);

$$E := \frac{2}{3}\frac{\sqrt{3}}{\cos(\theta)\sqrt{3} - \sin(\theta)}$$

> SumFx:=FE*sin(60*Pi/180)+DE*sin(60*Pi/180)+E*sin(theta);

$$SumFx := \frac{FE\sqrt{3}}{2} + \frac{DE\sqrt{3}}{2} + \frac{2}{3}\frac{\sqrt{3}\,\sin(\theta)}{\cos(\theta)\sqrt{3} - \sin(\theta)}$$

> SumFy:=FE*cos(60*Pi/180)-DE*cos(60*Pi/180)+E*cos(theta);

$$SumFy := \frac{FE}{2} - \frac{DE}{2} + \frac{2}{3}\frac{\sqrt{3}\,\cos(\theta)}{\cos(\theta)\sqrt{3} - \sin(\theta)}$$

> soln:=solve({SumFx=0,SumFy=0},{FE,DE});

$$soln := \{ DE = \frac{2}{3}, FE = -\frac{2}{3}\frac{\tan(\theta) + \sqrt{3}}{\sqrt{3} - \tan(\theta)} \}$$

> assign(soln);
> p:=plot([FE,DE,E],theta=0..Pi/4,color=black,labels=["theta (rads)",""],title="Non-dimensional Loads"):
> t1:=textplot([.7,.8,"DE/L"]):t2:=textplot([.7,-2.2,"FE/L"]):t3:=textplot([.7,1.9,"E/L"]):
> display(p,t1,t2,t3);

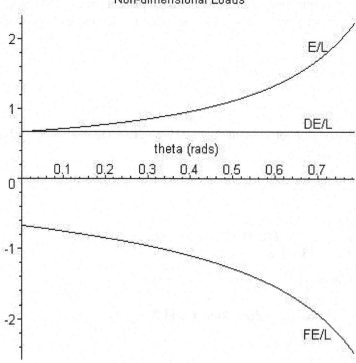

4.2 Problem 4/48 (Trusses. Method of Sections)

In this problem we would like to investigate the effects of the geometry of the upper part of the arched roof truss on some of the forces. We are given a constraint that the vertical distance between the horizontal section BF and the top most point must be 7 m, however we can vary the vertical location of points E and C. Determine the forces in members DE, EI, EF, FI, and HI of the truss. First, write the results in terms of the lengths a and b, then plot the forces as a function of a with the constraint that $a + b$ is always 7 m. Let a vary between 1 and 6 meters.

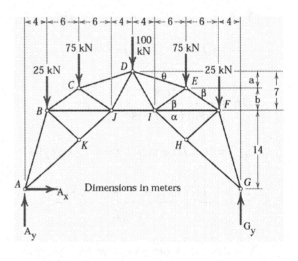

Problem Formulation

First we determine the reaction force at G from a free-body diagram for the entire truss (shown to the right). Due to symmetry,

$$A_y = G_y = (25 + 75 + 100 + 75 + 25)/2 = 150 \text{ kN}$$

From the free-body diagram we also find the angles,

$$\theta = \tan^{-1}\left(\frac{a}{6}\right) \quad \beta = \tan^{-1}\left(\frac{b}{6}\right) \quad \alpha = \tan^{-1}\left(\frac{14}{16}\right)$$

From the free-body diagram for section 1

$$\circlearrowleft \Sigma M_I = 0 = 150(16) - 25(12) + EF \sin \beta (12)$$

$$EF = -\frac{175}{\sin \beta}$$

$$\circlearrowleft \Sigma M_F = 0 = 150(4) - IH \sin \alpha (12)$$

$$IH = \frac{50}{\sin \alpha}$$

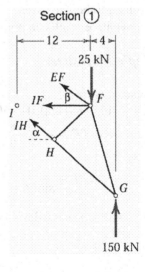

Section ①

150 kN

$$\Sigma F_x = 0 = -IF - EF \cos \beta - IH \cos \alpha$$

$$IF = -EF \cos \beta - IH \cos \alpha = \frac{175}{\tan \beta} - \frac{50}{\tan \alpha}$$

Note that the summation of moments about F was simplified by sliding IH along its line of action to I before resolving it into its horizontal and vertical components.

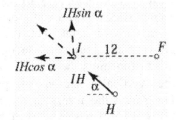

Now look at the free-body diagram for section 2. Forces IF and IH are already known, so we need only two equilibrium equations.

$$\Sigma M_I = 0 = DE \cos \theta (b) + DE \sin \theta (6)$$
$$- 75(6) - 25(12) + 150(16)$$

$$DE = \frac{-1650}{6 \sin \theta + b \cos \theta}$$

$$\Sigma F_x = 0 = IH \cos \alpha + IF + EI \cos \beta + DE \cos \theta$$

$$EI = -\frac{IH \cos \alpha + IF + DE \cos \theta}{\cos \beta}$$

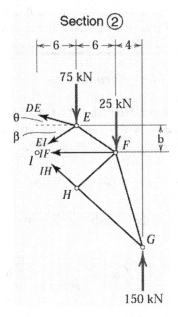

We will allow the computer to make the substitutions of IH, IF, and DE in the last equation. Note that all the members were assumed to be in tension. Thus positive results will imply tension and negative results compression.

Maple Worksheet

```
> restart; with(plots):
> EF:=-175/sin(beta);IH:=50/sin(alpha);
```

$$EF := -\frac{175}{\sin(\beta)}$$

$$IH := \frac{50}{\sin(\alpha)}$$

```
> IF:=-EF*cos(beta)-IH*cos(alpha);
```

$$IF := \frac{175 \, \cos(\beta)}{\sin(\beta)} - \frac{50 \, \cos(\alpha)}{\sin(\alpha)}$$

> DE:=-1650/(b*cos(theta)+6*sin(theta));

$$DE := -\frac{1650}{b \, \cos(\theta) + 6 \, \sin(\theta)}$$

> EI:=(-IH*cos(alpha)-IF-DE*cos(theta))/cos(beta);

$$EI := \frac{-\dfrac{175 \, \cos(\beta)}{\sin(\beta)} + \dfrac{1650 \, \cos(\theta)}{b \, \cos(\theta) + 6 \, \sin(\theta)}}{\cos(\beta)}$$

> theta:=arctan(a/6):beta:=arctan(b/6):alpha:=arctan(14/16):
> b:=7-a:
> p:=plot([EF,IH,IF,DE,EI],a=1..6,color=black,labels=["a (m)",""],title="Force kN"):
> t1:=textplot([5,-650,"EF"]):t2:=textplot([5.5,150,"IH"]):t3:=textplot([5,550,"IF"]):
> t4:=textplot([5.8,-250,"DE"]):t5:=textplot([5.8,-550,"EI"]):
> display(p,t1,t2,t3,t4,t5);

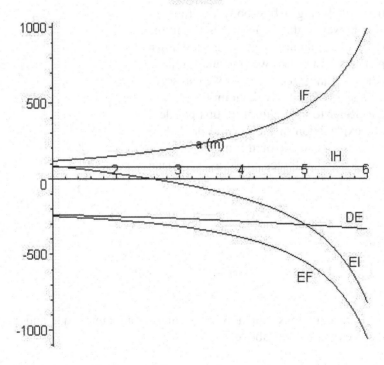

4.3 Sample Problem 4/5 (Space Trusses)

The space truss consists of the rigid tetrahedron *ABCD* anchored by a ball-and-socket connection at *A* and prevented from any rotation about the *x*-, *y*-, or *z*-axes by the respective links 1, 2, and 3. The load *L* is applied at joint *E*, which is rigidly fixed to the tetrahedron by the three additional links. Here we would like to investigate how the forces in a few of the members depend upon the height of the structure so let the height be *d* instead of 4 meters. Solve for the forces in the members at joint *E* and plot these as a function of *d* if *L* = 10 kN. Let *d* range between 2 and 10 meters.

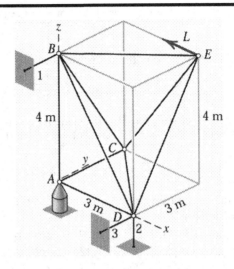

Problem Formulation

We could begin by analyzing a free-body diagram for the entire structure; however, this is not necessary in the present case as joint *E* contains only three unknown forces that happen to be the ones we are interested in. Joint *E* also has the external force *L* so that we can solve this problem from a single free-body diagram for joint *E*, shown to the right. Refer to the solution to this problem in your text for an explanation of the general procedure for determining the forces in all of the members. Our first step is to express the forces at *E* as Cartesian vectors.

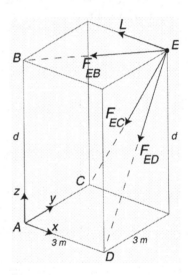

$$\mathbf{F_{EC}} = F_{EC}\frac{-3\mathbf{i}-d\mathbf{k}}{\sqrt{9+d^2}} \qquad \mathbf{F_{ED}} = F_{ED}\frac{-3\mathbf{j}-d\mathbf{k}}{\sqrt{9+d^2}}$$

$$\mathbf{F_{EB}} = \frac{F_{EB}}{\sqrt{2}}(-\mathbf{i}-\mathbf{j}) \qquad \mathbf{L} = -L\mathbf{i}$$

The scalar equilibrium equations can now be written from the *x*-, *y*-, and *z*-components of the vector equations above.

$$\Sigma F_x = 0 = -L - \frac{F_{EB}}{\sqrt{2}} - \frac{3F_{EC}}{\sqrt{9+d^2}} \qquad\qquad \Sigma F_y = 0 = -\frac{F_{EB}}{\sqrt{2}} - \frac{3F_{ED}}{\sqrt{9+d^2}}$$

$$\Sigma F_z = 0 = -\frac{d}{\sqrt{9+d^2}}\left(F_{EC} + F_{ED}\right)$$

These equations can be solved for the three forces,

$$F_{ED} = -F_{EC} = \frac{L}{6}\sqrt{9+d^2} \qquad F_{EB} - \frac{L}{\sqrt{2}}$$

Referring back to the free-body diagram we see that all three members were assumed to be in tension. Thus, the results indicate that *ED* is in tension while *EC* and *EB* are in compression.

Maple Worksheet

> restart; with(plots):
> eqn1:=-L-Feb/sqrt(2)-3*Fec/sqrt(9+d^2)=0;

$$eqn1 := -L - \frac{1}{2}Feb\sqrt{2} - \frac{3\,Fec}{\sqrt{9+d^2}} = 0$$

> eqn2:=-Feb/sqrt(2)-3*Fed/sqrt(9+d^2)=0;

$$eqn2 := -\frac{1}{2}Feb\sqrt{2} - \frac{3\,Fed}{\sqrt{9+d^2}} = 0$$

> eqn3:=Fec+Fed=0;

$$eqn3 := Fec + Fed = 0$$

> soln:=solve({eqn1,eqn2,eqn3},{Feb,Fec,Fed});

$$soln := \{ Fec = -\frac{1}{6}L\sqrt{9+d^2},\ Feb = -\frac{1}{2}L\sqrt{2},\ Fed = \frac{1}{6}L\sqrt{9+d^2}\}$$

> assign(soln);
> L:=10:
> t1:=textplot([8,16,"F_ed"]): t2:=textplot([8,-5.3,"F_eb"]):
  t3:=textplot([8,-12,"F_ec"]):
> p:=plot([Feb,Fec,Fed],d=2..10,labels=["d (m)",""], title="Force (kN)"):
> display(p,t1,t2,t3);

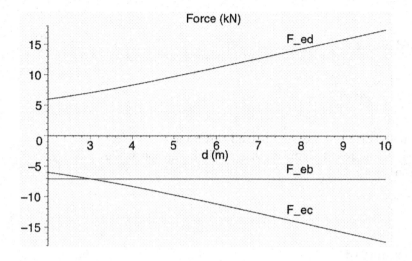

4.4 Problem 4/83 (Frames and Machines)

Determine the magnitude of the pin reaction at A and the magnitude of the force reaction at the rollers as a function of x. The pulleys at C and D are small. Plot the two forces for $0 \le x \le 0.8$ m. What is the direction of the force at the rollers? Does the direction change for the range of x considered?

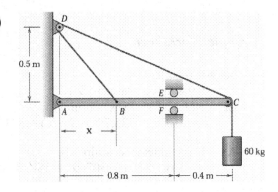

Problem Formulation

$$\alpha = \tan^{-1}(0.5/x) \qquad \theta = \tan^{-1}(0.5/1.2) = 22.62°$$

$$\circlearrowleft M_A = 0 = Tx\sin\alpha + T(1.2)\sin\theta - T(1.2) + EF(0.8)$$

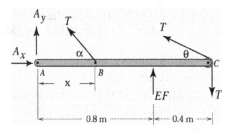

Note in the Free Body Diagram that we have assumed that force EF acts up. This is equivalent to assuming that the boom is in contact with and pushing down on roller F. Roller F in turn pushes back up on the boom. This assumption is verified by calculating only positive values for the force.

Substituting $\theta = 22.62°$, T = 60(9.81), and solving gives,

$$EF = 543.32 - 735.75x\sin\alpha$$

$$\Sigma F_x = 0 = A_x - T\cos\theta - T\cos\alpha \qquad A_x = T(\cos\theta + \cos\alpha)$$

$$\Sigma F_y = 0 = A_y + T\sin\alpha + T\sin\theta + EF - T \qquad A_y = T(1 - \sin\alpha - \sin\theta) - EF$$

$$A = \sqrt{A_x^2 + A_y^2}$$

The formulation is now complete since the tow forces EF and A are known as a function of x. Note once again that we do not have to make explicit substitutions. For example, EF is written in terms of x and α, but α is already known as a function of x.

Maple Worksheet

```
> restart; with(plots):
> T:=60*9.81:theta:=arctan(.5/1.2):
> EF:= 543.32-735.75*x*sin(alpha);
```
$$EF := -735.75 \ x \sin(\alpha) + 543.32$$

```
> Ax:=T*(cos(theta)+cos(alpha));
```
$$Ax := 543.3230769 \ + 588.60 \ \cos(\alpha)$$

```
> Ay:=T*(1-sin(theta)-sin(alpha))-EF;
```
$$Ay := -181.1046154 \ - 588.60 \ \sin(\alpha) + 735.75 \ x \sin(\alpha)$$

```
> A:=sqrt(Ax^2+Ay^2):
> alpha:=arctan(.5/x):
> p:=plot([EF,A],x=0..0.8,color=black,labels=["x (m)",""],title="Force (N)"):
> t1:=textplot([.7,300,"EF"]):t2:=textplot([.7,1000,"A"]):
> display(p,t1,t2);
```

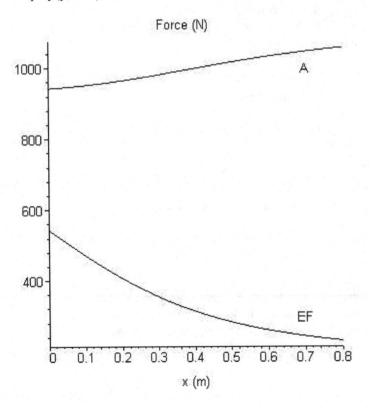

4.5 Problem 4/108 (Frames and Machines)

A lifting device for transporting 135-kg steel drums is shown. Develop expressions for the forces at A, C, and E in terms of L, the length of member AG. Plot these forces as a function L between 345 and 380 mm.

Problem Formulation

$$\theta = \cos^{-1}(340/L) \qquad P = W = 135(9.81) \text{ N}$$

Since BG and AG are two-force members we can draw a free-body of pin G as shown to the right. By symmetry, $AG = BG$, so

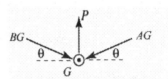

$$\Sigma F_y = 0 = P - 2AG\sin\theta$$

$$AG = \frac{P}{2\sin\theta}$$

Now consider the free-body diagram of ACE.

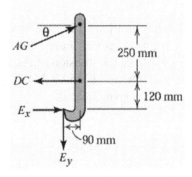

$$\circlearrowleft M_E = 0 = AG\sin\theta(90) - AG\cos\theta(370) + DC(120)$$

$$DC = \left(\frac{37}{12}\cos\theta - \frac{3}{4}\sin\theta\right)AG$$

$$\Sigma F_x = 0 = E_x - DC + AG\cos\theta \qquad E_x = DC - AG\cos\theta$$

$$\Sigma F_y = 0 = AG\sin\theta - E_y \qquad E_y = AG\sin\theta = \frac{P}{2}$$

$$E = \sqrt{E_x^2 + E_y^2}$$

The formulation is now complete since all forces are known in terms of θ which is in turn known as a function of L.

Maple Worksheet

```
> restart; with(plots):
> theta:=arccos(340/L):
> AG:=P/2/sin(theta);
```

$$AG := \frac{P}{2\sqrt{1 - \dfrac{115600}{L^2}}}$$

```
> DC:=37/12*AG*cos(theta)-9/12*AG*sin(theta);
```

$$DC := \frac{3145\ P}{6\sqrt{1 - \dfrac{115600}{L^2}}\ L} - \frac{3\ P}{8}$$

```
> Ex:=DC-AG*cos(theta);
```

$$Ex := \frac{2125\ P}{6\sqrt{1 - \dfrac{115600}{L^2}}\ L} - \frac{3\ P}{8}$$

```
> E:=sqrt(Ex^2+Ey^2):
> P:=135*9.81:Ey:=P/2:
>
> p:=plot([AG/1000,DC/1000,E/1000],L=345...380,color=black,labels=["L
(mm)",""],title="Force (kN)"):
> # Note the division of the forces by 1000 to yield kN
> t1:=textplot([350,9.2,"DC"]):t2:=textplot([350,6.0,"E"]):t3:=textplot([350,3.4,"AG"]):
> display(p,t1,t2,t3);
```

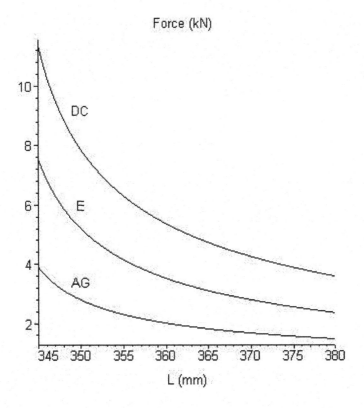

DISTRIBUTED FORCES

5

This chapter considers forces that are distributed over lines, areas, or volumes. Also considered are a few preliminary topics directly related to distributed forces such as center of mass and centroids. The major applications are beams, cables and fluid statics.

Problem 5.1 contains the first example in this booklet of a parametric plot. Problem 5.2 formulates the equilibrium equations for a beam in terms of an integral of the distributed load on the beam. Maple's *int* command is used to evaluate the integrals. Problem 5.3 expresses the internal shear force and moment in terms of integrals of the external loading and then uses Maple's *int* to evaluate the integrals. *fsolve* is used to obtain a numerical solution for the minimum tension and total length of a flexible cable in problem 5.4. Problem 5.4 also illustrates the use of a do loop in Maple.

Following is a summary of the Maple commands used in this chapter.

Table 5.1 Maple Commands used in Chapter 5

Maple Command	Problem
for, do, end do	5.4
fsolve	5.4
display	5.1, 5.2, 5.3, 5.4
display(*seq*)	5.4
evalf	5.3
int	5.2, 5.3
plot (normal)	5.2, 5.3, 5.5
plot (parametric)	5.1
simplify	5.2
solve	5.2, 5.5
subs	5.2, 5.3
textplot	5.1, 5.2, 5.3, 5.4, 5.5

5.1 Sample Problem 5/8 (Composite Bodies)

Let the length of the 40-mm diameter shaft in the bracket-and-shaft assembly be L (mm). The vertical face of the assembly is made from sheet metal that has a mass of 25 kg/m². The material of the horizontal base has a mass of 40 kg/m², and the steel shaft has a density of 7.83 Mg/m³. (a) Find expressions for the location of the center of mass ($\overline{Y}$, $\overline{Z}$) in terms of L. (b) Plot $\overline{Y}$ and $\overline{Z}$ as a function of L for $0 \le L \le 200$ mm. (c) Plot $\overline{Z}$ versus $\overline{Y}$ for $0 \le L \le 200$ mm.

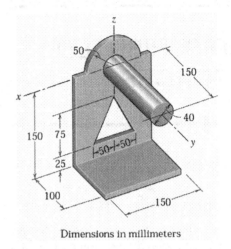

Dimensions in millimeters

Problem Formulation

(a) Refer to the solution to this sample problem in your text and be sure that you understand how the results in the table were obtained. For the problem considered here, the table will remain unchanged except for the last row. For the steel shaft (part 5) we have,

$$m = 0.00984L; \quad \overline{y} = L/2; \quad \overline{z} = 0; \quad m\overline{y} = 0.00492L^2; \quad m\overline{z} = 0.$$

The totals become

$$\Sigma m = 1.166 + 0.00984L; \quad \Sigma m\overline{y} = 30 + 0.00492L^2; \quad \Sigma m\overline{z} = -120.73$$

Thus,

$$\overline{Y} = \frac{\Sigma m\overline{y}}{\Sigma m} = \frac{30 + 0.00492L^2}{1.166 + 0.00984L}; \quad \overline{Z} = \frac{\Sigma m\overline{z}}{\Sigma m} = \frac{-120.73}{1.166 + 0.00984L}$$

(b) The plot of $\overline{Y}$ and $\overline{Z}$ versus L can be found in the Maple worksheet below.

(c) At first sight, plotting $\overline{Z}$ versus $\overline{Y}$ may seem problematic since $\overline{Z}$ is not known explicitly as a function of $\overline{Y}$. Perhaps the most straightforward way of dealing with this apparent difficulty is to solve the first equation above for L in terms of $\overline{Y}$ and then substitute this result into the second equation. Though this approach might seem simple at first, it turns out that there are two solutions, both of which are very messy. There may also be situations where this substitution approach may be impossible, even with symbolic algebra.

Fortunately, most software packages allow plotting two variables without performing a substitution, provided the two variables are expressed in terms of a common parameter (in our case L). Plots of this type are usually called parametric plots. The parametric plot of $\bar{Z}$ versus $\bar{Y}$ can be found in the Maple worksheet below.

Maple Worksheet

```
> restart; with(plots):
> sum_mass:=1.166+0.00984*L;
```
$$sum\_mass := 1.166 + .00984\,L$$

```
> yb:=(30+0.00492*L^2)/sum_mass;
```
$$yb := \frac{30 + .00492\,L^2}{1.166 + .00984\,L}$$

```
> zb:=-120.73/sum_mass;
```
$$zb := -120.73\,\frac{1}{1.166 + .00984\,L}$$

```
> p1:=plot([yb,zb],L=0..200,labels=[`L (mm)`,` (mm)`]):
> t1:=textplot([180,73,`Y`]): t2:=textplot([180,-33,`Z`]):
> display(p1,t1,t2);
```

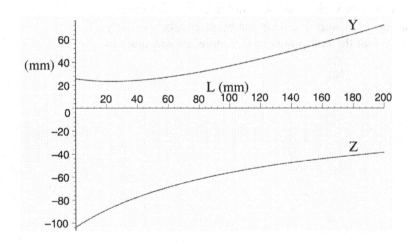

```
> p2:=plot([yb,zb,L=0..200],x=0..80,y=-110..0,labels=[`y (mm)`,`z
(mm)`],labeldirections=[HORIZONTAL,VERTICAL]):
> t3:=textplot([25,-110,`L = 0`]): t4:=textplot([70,-30,`L = 200 mm`]):
> display(p2,t3,t4);
```

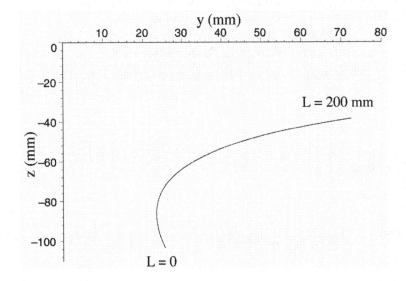

In this problem we have seen two ways of presenting the location of the mass center of the bracket-and-shaft assembly and it is useful to compare the advantages and disadvantages of these methods. The first plot shows clearly how $\overline{Y}$ and $\overline{Z}$ depend on L, however, it is difficult to picture the spatial location of the mass center. In the second plot one can actually see the spatial location of the mass center and how it moves as L is varied. What is not seen are the precise values of L at each combination of $\overline{Y}$ and $\overline{Z}$. This situation can be partially remedied by giving the values of L at the end-points of the curve, as was done in the plot above.

5.2 Problem 5/113 (Beams-External Effects)

The load per foot of beam length varies as shown. Here we would like to look at the effects of the initial load w_0 on the reactions given the constraint that the total resultant R of the distributed load w remains constant. This means that we are, in effect, studying the effects of the shape of the load. First find the required value of k in terms of w_0 and L and then (a) plot the distributed load $(w(x))$ for $w_0 = 100, 200,$ and 300 lb/ft and (b) plot the reactions at the two supports as a function of w_0 for $0 \le w_0 \le 300$ lb/ft. For (a) and (b) let $R = 3500$ lb and $L = 20$ ft.

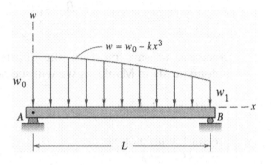

Problem Formulation

$$R = \int_0^L w\,dx = \int_0^L \left(w_0 - kx^3\right)dx = w_0 L - \frac{1}{4}kL^4$$

Solving,

$$k = \frac{4\left(w_0 L - R\right)}{L^4}$$

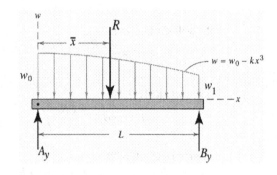

$$\bar{x} = \frac{1}{R}\int_0^L xw\,dx = \frac{1}{R}\int_0^L x\left(w_0 - kx^3\right)dx = \frac{1}{R}\left(\frac{1}{2}w_0 L^2 - \frac{1}{5}kL^5\right)$$

Substituting for k,

$$\bar{x} = \frac{L\left(8R - 3w_0 L\right)}{10R}$$

Now, from the free-body diagram,

$$\circlearrowleft \Sigma M_A = 0 = B_y L - R\bar{x}$$

$$\Sigma F_y = 0 = A_y + B_y - R$$

From which we find,

$$B_y = \frac{\bar{x}}{L} R \qquad A_y = R - B_y = \left(1 - \frac{\bar{x}}{L}\right) R$$

Needless to say, most of the substitutions indicated above are performed automatically in Maple, as we will see below.

Maple Worksheet

> restart; with(plots):
> w:=w0-k*x^3;

$$w := w0 - k\,x^3$$

> eqn1:=R=int(w,x=0..L);

$$eqn1 := R = w0\,L - \frac{1}{4}\,k\,L^4$$

> k:=solve(eqn1,k);

$$k := -\frac{4\,(R - w0\,L)}{L^4}$$

> w1:=w0-k*L^3;

$$w1 := w0 + \frac{4\,(R - w0\,L)}{L}$$

> xbar:=1/R*int(w*x,x=0..L);

$$xbar := \frac{\dfrac{4\,(R - w0\,L)\,L}{5} + \dfrac{w0\,L^2}{2}}{R}$$

> simplify(%);

$$\frac{L\,(8\,R - 3\,w0\,L)}{10\,R}$$

> R:=3500;L:=20;

$$R := 3500$$

$$L := 20$$

> B:=xbar*R/L;

$$B := 2800 - 6\,w0$$

> A:=R-B;

$$A := 700 + 6\,w0$$

> wa:=subs(w0=100,w):wb:=subs(w0=200,w):wc:=subs(w0=300,w):

> p1:=plot([wa,wb,wc],x=0..L,color=black,labels=["x (ft)",""],title="Load w (lb/ft)"):
> display(p1);

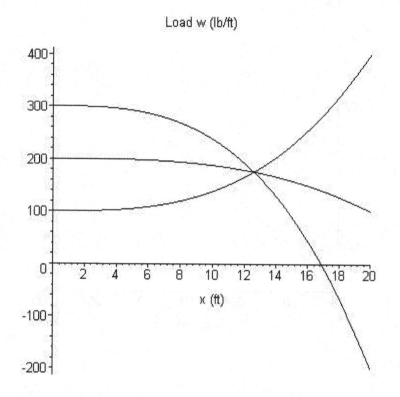

>
> p2:=plot([A,B],w0=0..300,color=black,labels=["w0 (lb/ft)",""],title="Support Reactions (lb)"):
> t1:=textplot([250,2300,"A"]):t2:=textplot([250,1200,"B"]):
> display(p2,t1,t2);

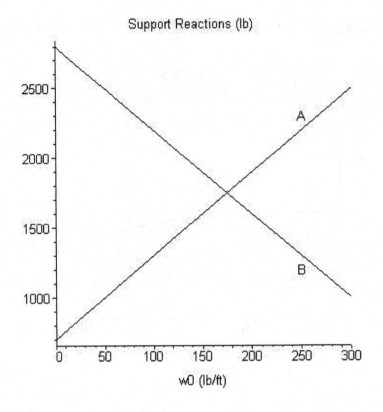

5.3 Problem 5/140 (Beams-Internal Effects)

Derive expressions for the shear force V and bending moment M as functions of x for the cantilever beam loaded as shown. For the case where $L = 10$ ft and $w_L = 400$ lb/ft, (a) plot the distributed load $w(x)$, (b) the shear force $V(x)$ and (c) the bending moment $M(x)$. In each case plot three curves for $w_0 = -400, 0,$ and 400 lb/ft.

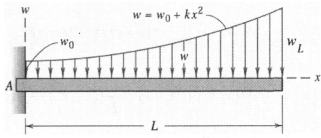

Problem Formulation

First, we need to express k in terms of w_0 and w_L. This is accomplished by substituting $x = L$ into the loading function,

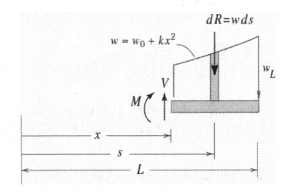

$$w(x = L) = w_0 + kL^2 = w_L$$

$$k = \frac{w_L - w_0}{L^2}$$

The easiest way to find $V(x)$ and $M(x)$ is from the free-body diagram shown to the right since we avoid having to find the reactions at the wall support. We do, however, have to be very careful setting up our integrals.

When we say $V(x)$, x is a coordinate measured positive to the right from the left end of the beam. It is essential to distinguish between that x and the dummy integration variable which will vary between x and L. Here, we denote the integration variable by s.

$$\Sigma F_y = V - \int dR \qquad V = \int_x^L w(s)\,ds = \int_x^L \left(w_0 + ks^2\right)ds$$

$$V(x) = w_0(L - x) + \frac{k}{3}\left(L^3 - x^3\right)$$

Before we sum moments first note that the moment arm of dR about a point on the cross-section at x is $s - x$,

$$\circlearrowleft \Sigma M = 0 = -M - \int (s-x)dR$$

$$M(x) = -\int_{x}^{L}(s-x)w(s)ds = -\int_{x}^{L}(s-x)\left(w_0 + ks^2\right)ds$$

$$M(x) = xw_0(L-x) - \frac{w_0}{2}\left(L^2 - x^2\right) + \frac{xk}{3}\left(L^3 - x^3\right) - \frac{k}{4}\left(L^4 - x^4\right)$$

Maple Worksheet

```
> restart; with(plots):
> w:=w0+k*x^2;
```
$$w := w0 + k\,x^2$$

```
> V:=int(w0+k*s^2,s=x..L);
```
$$V := w0\,(L-x) + \frac{k\,(L^3 - x^3)}{3}$$

```
> M:=-int((s-x)*(w0+k*s^2),s=x..L);
```
$$M := -\frac{k\,(L^4 - x^4)}{4} + \frac{x\,k\,(L^3 - x^3)}{3} - \frac{w0\,(L^2 - x^2)}{2} + x\,w0\,(L-x)$$

```
> L:=10;wL:=400;k:=(wL-w0)/L^2;
```
$$L := 10$$

$$wL := 400$$

$$k := 4 - \frac{w0}{100}$$

```
> wa:=subs(w0=-400,w): wb:=subs(w0=0,w): wc:=subs(w0=400,w):
> p1:=plot([wa,wb,wc],x=0..L,color=black,labels=["x (ft)",""],title="Load w (lb/ft)"):
>
> display(p1);
```

Load w (lb/ft)

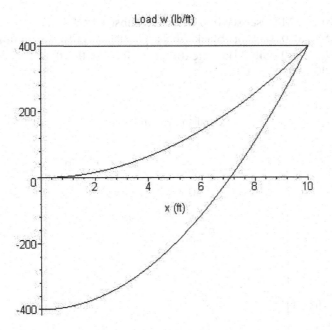

```
>
> Va:=subs(w0=-400,V): Vb:=subs(w0=0,V): Vc:=subs(w0=400,V):
> p2:=plot([Va,Vb,Vc],x=0..L,color=black,labels=["x (ft)",""],title="Shear Force V (lb)"):
> t1:=textplot([3,3300,"w0 = 400"]):t2:=textplot([3,1500,"w0 = 0"]):
> t3:=textplot([2.5,-800,"w0 = -400"]):
> display(p2,t1,t2,t3);
```

Shear Force V (lb)

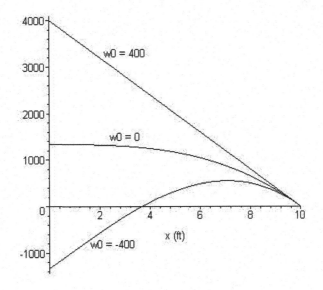

> Ma:=subs(w0=-400,M): Mb:=subs(w0=0,M): Mc:=subs(w0=400,M):
> p3:=plot([Ma,Mb,Mc],x=0..L,color=black,labels=["x (ft)",""],title="Bending Moment (lb-ft)"):
> t4:=textplot([3,-13000,"w0 = 400"]):t5:=textplot([3,-4600,"w0 = 0"]):
> t6:=textplot([4,-1600,"w0 = -400"]):
> display(p3,t4,t5,t6);

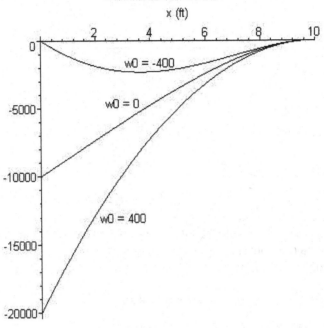

5.4 Sample Problem 5/17 (Flexible Cables)

Replace the cable of Sample Problem 5/16, which is loaded uniformly along the horizontal, by a cable which has a mass of 12 kg per meter of its own length and supports its own weight only. The cable is suspended between two points on the same level 300 m apart and has a sag of h meters (instead of the 60 m shown in the diagram. (a) Plot y as a function of x for $h = 10$, 30, and 60 meters. (b) Plot the minimum tension T_0 and the total length ($L_c = 2s$) of the cable as a function of the sag h.

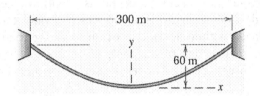

Problem Formulation

(a) For a uniformly distributed load we have a catenary shape for the cable as described in part (c) of Article 5/8. The curve assumed by the cable ($y(x)$) is thus described by Equation 5/19

$$ y = \frac{T_0}{\mu}\left(\cosh\frac{\mu x}{T_0} - 1\right) $$

where $\mu = (12)(9.81)(10^{-3}) = 0.1177$ kN/m. Before y can be plotted as a function of x it is first necessary to find the value of T_0 corresponding to the three values of h that are given. As in the sample problem, this is accomplished by evaluating the above equation at $x = 150$ m,

$$ h = \frac{T_0}{\mu}\left(\cosh\frac{150\mu}{T_0} - 1\right) $$

Since there are only three cases to consider it might seem reasonable to try the graphical method used in the sample problem, however, this approach is obviously too cumbersome for part (b). Therefore it is best to use a numerical approach. This is easily accomplished using Maple's *fsolve* function. The results are $T_0 = 132.6$, 44.7, and 23.2 kN for $h = 10$, 30, and 60 meters respectively. With T_0 known y can easily be plotted as a function of x by substituting into the equation above. The result is given in the worksheet below.

(b) It is possible to use the general approach for part (a) here as well. What you would need to do is set up a table of values for h and T_0. You would then use *fsolve* to find T_0 for each value of h in the table. Finally, the tabular results would be entered into a graphics program (or a spreadsheet such as Excel) and the results plotted.

It turns out that you can automate the process with only a few statements in Maple. The general procedure is as follows (see the worksheet for details). First, a do-loop structure is set up which iterates on *h*. The Maple command *fsolve* is placed within the loop to evaluate T_0 for each *h*. After T_0 has been determined, the total length of the cable ($L_c = 2s$) can be easily found from Equation 5/20 (with $x = 150$ m). The results are then plotted after the loop has completed.

Maple Worksheet

> restart;with(plots):with(plottools):

First we set up Equation 5/19. It is very convenient for plotting to define y explicitly as a function of T0.

> y:=T_0->T_0/mu*(cosh(mu*x/T_0)-1);

$$y := T\_0 \rightarrow \frac{T\_0\left(\cosh\left(\frac{\mu x}{T\_0}\right) - 1\right)}{\mu}$$

Now set up another equation which evaluates y = h at x = L.

> eqn1:=h=T0/mu*(cosh(mu*L/2/T0)-1);

$$eqn1 := h = \frac{T0\left(\cosh\left(\frac{1}{2}\frac{\mu L}{T0}\right) - 1\right)}{\mu}$$

> L:=300: mu:=0.1177:

Now we solve eqn1 for the three cases of h. We'll denote the values of T0 as T0_10, T0_30, and T0_60 for h = 10, 30, and 60 meters respectively.

> T0_10:=fsolve(subs(h=10,eqn1),T0);

$$T0\_10 := 132.6082037$$

> T0_30:=fsolve(subs(h=30,eqn1),T0);

$$T0\_30 := 44.71391316$$

> T0_60:=fsolve(subs(h=60,eqn1),T0);

$$T0\_60 := 23.15850794$$

Now we can plot the function y defined above for these three cases. Note that we use "scaling=constrained" to get the x and y axes to plot to scale.

> t1:=textplot([150,60,"60"]):

```
> t2:=textplot([150,30,"30"]):
> t3:=textplot([150,10,"10"]):
> c1:=plot([y(T0_10),y(T0_30),y(T0_60)],x = -L/2..L/2, scaling=constrained ,
labels=[`x (m)`,`y (m)`]):
> display(c1,t1,t2,t3);
```

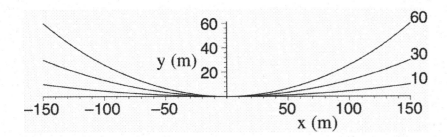

As discussed in the problem formulation, the easiest way to obtain the plots for part (b) is to set up a do loop. Note that the force T0 and cable length Lc are both calculated within the loop. Also note that plot structures (data points p1 and p2) are defined within the loop as well. This is very convenient but requires using a different sort of plotting than was discussed in Chapter 1. Basically, what we do is plot the sequence of points determined in the loop. The disadvantage of this approach is that labels are not added with the *display(seq)* command. The labels in the plots below were added with a graphics package.

```
> for i from 5 to 60 do
h:=i:
T0[i]:=fsolve(eqn1,T0):
p1[i]:=point([h,T0[i]], symbol=circle):
Lc[i]:=2*T0[i]/mu*sinh(mu*L/2/T0[i]): # cable length =2s (Eqn 5/20)
p2[i]:=point([h,Lc[i]], symbol=circle):
end do:
> display(seq(p1[i],i=5..60));
```

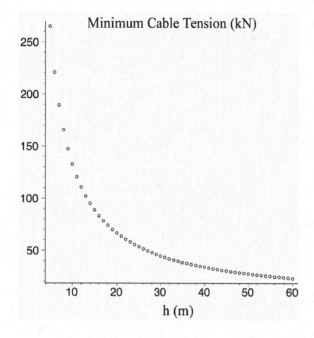

> display(seq(p2[i],i=5..60));

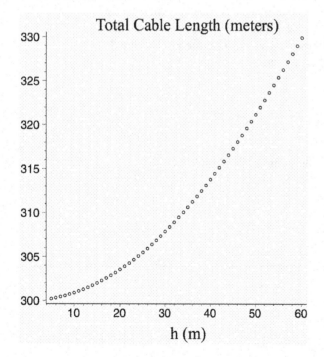

5.5 Problem 5/189 (Fluid Statics)

The rectangular gate shown in section is 10 ft long (perpendicular to the paper) and is hinged about its upper edge B. The gate divides a channel leading to a fresh-water lake on the left and a salt-water tidal basin on the right. Calculate the torque M on the shaft of the gate at B required to prevent the gate from opening in terms of h (the distance between salt water and fresh water levels). Plot M as a function of h for $0 \le h \le 6$ ft.

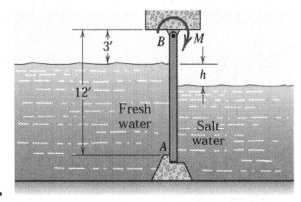

Problem Formulation

To the right is a free-body diagram for the gate. The water pressures have been replaced by linear load distributions where the maximum load intensity w is given by the general formula

$$w = \gamma dL$$

where γ is the specific weight, d is the depth of water and L is the length (into the page). Thus,

$$w_f = \gamma_f (9\,ft)(10\,ft) \qquad w_s = \gamma_s (9 - h)(10)$$

where $\gamma_f = 62.4 \dfrac{lb}{ft^3}$ and $\gamma_s = 64.0 \dfrac{lb}{ft^3}$.

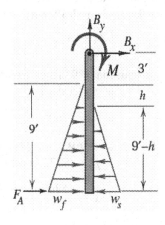

Now replace the distributed loads by their statically equivalent concentrated loads as shown on the free-body diagram to the right. Note that the contact force $F_A = 0$ since we are finding the smallest M required, i.e. the gate is on the verge of opening.

$$\circlearrowleft \Sigma M_B = 0 = P_f h_1 - P_s h_2 - M$$

$$M = P_f h_1 - P_s h_2$$

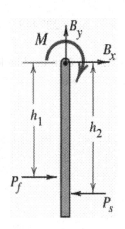

where,

$$h_1 = 3 + \frac{2}{3}(9) = 9\,ft \qquad\qquad h_2 = 3 + h + \frac{2}{3}(9-h) = 9 + \frac{h}{3}$$

$$P_f = \frac{1}{2}w_f(9) = \frac{1}{2}\gamma_f(9)^2(10)$$

$$P_s = \frac{1}{2}w_s(9-h) = \frac{1}{2}\gamma_s(9-h)^2(10)$$

Maple Worksheet

```
> restart; with(plots):
> gamma_f:=62.4:gamma_s:=64:
> h1:=9:h2:=3+h+2/3*(9-h);
```
$$h2 := 9 + \frac{h}{3}$$

```
> w_f:=gamma_f*9*10;w_s:=gamma_s*(9-h)*10;
```
$$w\_f := 5616.0$$

$$w\_s := 5760 - 640\,h$$

```
> P_f:=1/2*w_f*9;P_s:=1/2*w_s*(9-h);
```
$$P\_f := 25272.00000$$

$$P\_s := \frac{(5760 - 640\,h)(9-h)}{2}$$

```
> M:=P_f*h1-P_s*h2;
```
$$M := 227448.0000 - \frac{(5760 - 640\,h)(9-h)\left(9 + \frac{h}{3}\right)}{2}$$

```
> plot(M,h=0..6,color=black,labels=["h (ft)",""],title="Moment (lb-ft)");
```

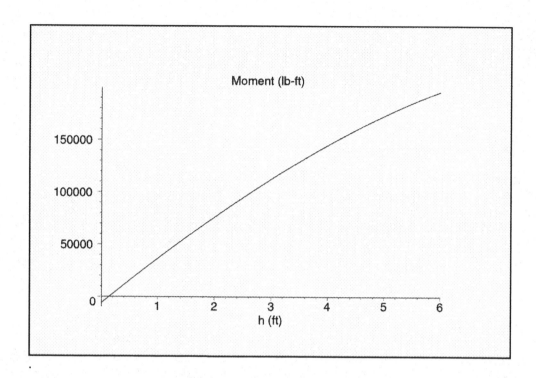

FRICTION

6

Coulomb friction (dry friction) can have a significant effect upon the analysis of engineering structures. Problem 6.1 considers three blocks stacked on an incline oriented at an arbitrary angle θ and illustrates the importance of identifying all possibilities for impending motion. In this problem there turns out to be a transition between two possibilities as the angle θ increases. Problem 6.2 considers impending slip of a cylinder wedged between two rough surfaces. Maple's *solve* command is used to solve three equations symbolically. Problem 6.3 considers the possibility of impending slip as a person climbs to the top of a ladder. Problems 6.4, 6.5, and 6.6 are applications involving friction on wedges, screws and flexible belts.

Following is a summary of the Maple commands used in this chapter.

Table 6.1 Maple Commands used in Chapter 6

Maple Command	Problem
display	6.1, 6.2, 6.3, 6.4, 6.5, 6.6
plot (normal)	6.1, 6.2, 6.3, 6.4, 6.5, 6.6
solve	6.2, 6.3, 6.4
subs	6.2, 6.3, 6.4
textplot	6.1, 6.2, 6.3, 6.4, 6.5, 6.6

6.1 Sample Problem 6/5 (Friction)

The three flat blocks are positioned on an incline that is oriented at an angle θ (instead of 30°) from the horizontal. Plot the maximum value of P (if no slipping occurs) versus θ. Consider only positive values of P and indicate the regions over which (1) the 50-kg block slides alone and (2) the 50-kg and 40-kg blocks slide together. The coefficients of friction μ_s are given in the figure to the right.

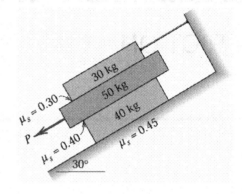

Problem Formulation

The free-body diagrams for the three blocks are shown to the right. To obtain the required plot it will be convenient to use a slightly different approach than that used in the sample problem in your text. We start by writing down the equilibrium equations without making any assumptions about where sliding occurs.

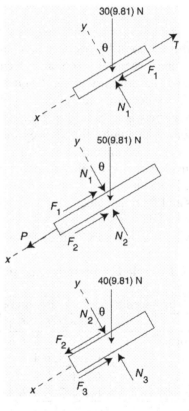

$$\left[\Sigma F_y = 0\right] \quad \text{(30-kg)} \quad N_1 - 30(9.81)\cos\theta = 0$$
$$\text{(50-kg)} \quad N_2 - N_1 - 50(9.81)\cos\theta = 0$$
$$\text{(40-kg)} \quad N_3 - N_2 - 40(9.81)\cos\theta = 0$$

These equations can be readily solved for the normal forces.

$$N_1 = 30(9.81)\cos\theta \quad N_2 = 80(9.81)\cos\theta$$
$$N_3 = 120(9.81)\cos\theta$$

$$\left[\Sigma F_x = 0\right] \quad \text{(50-kg)} \quad P - F_1 - F_2 + 50(9.81)\sin\theta = 0$$
$$\text{(40-kg)} \quad F_2 - F_3 + 40(9.81)\sin\theta = 0$$

Now we have two equations with four unknowns, P, F_1, F_2, and F_3. Note that we have not written the equation for the summation of forces in the x direction for the 30-kg block. The reason is that this equation introduces an additional unknown (T) that we are not interested in determining.

The next step is to make assumptions about which block(s) slide. As will be seen, either of the two possible assumptions about impending motion will reduce two of the friction forces to functions of θ only. This will result in two equations that may be solved for P and the remaining friction force. The forces calculated will be designated P_1 or P_2 to distinguish the two cases for impending slip.

Case (1): Only the 50-kg block slips.

Impending slippage at both surfaces of the 50-kg block gives $F_1 = 0.3N_1 = 88.29\cos\theta$ and $F_2 = 0.4N_1 = 313.9\cos\theta$. Substituting these results into the equilibrium equations yields

$$P_1 = 402.2\cos\theta - 490.5\sin\theta$$

Case (2): The 40 and 50-kg blocks slide together.

Impending slippage at the upper surface of the 40-kg block and lower surface of the 50-kg block gives $F_1 = 0.3N_1 = 88.29\cos\theta$ and $F_3 = 0.45N_3 = 529.7\cos\theta$. Substitution of these results into the equilibrium equations gives,

$$P_2 = 618.0\cos\theta - 882.9\sin\theta$$

Which of these two values of P represents the maximum load that can be applied without slippage on any surface is best illustrated by plotting the two expressions as a function of θ. This plot will be generated in the worksheet below. The basic idea is that at any specified angle θ, the critical or maximum value of P will be the smaller of two values calculated.

Maple Worksheet

```
> restart; with(plots):with(plottools):
> P1:=402.2*cos(theta)-490.5*sin(theta);
        P1 := 402.2 cos(θ) − 490.5 sin(θ)

> P2:=618.0*cos(theta)-882.9*sin(theta);
        P2 := 618.0 cos(θ) − 882.9 sin(θ)

> p1:=plot([P1,P2],theta=0..Pi/4,y=0..620, labels=["theta (rads)","P (N)"],
labeldirections=[HORIZONTAL,VERTICAL]):
> t1:=textplot([0.1,390,"P1"]):
> t2:=textplot([0.1,580,"P2"]):
> display(p1,t1,t2);
```

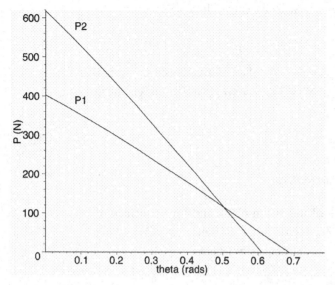

The figure above shows P_1 and P_2 plotted as a function of θ. For each θ, the critical or maximum value of P will be the smaller of two values calculated. By setting $P_1 = P_2$ we find that the two curves intersect at $\theta = 0.503$ rads (28.8°). Thus, for $\theta \leq 28.8°$ P_1 controls and the 50-kg block slides by itself while for $\theta \geq 28.8°$ P_2 controls and the 40 and 50-kg blocks slide together.

The figure below shows only the critical values for P together with an indication of the mode of slippage. This type of figure is rather tedious to produce in Maple so the code is not given. It is shown here for purposes of illustration.

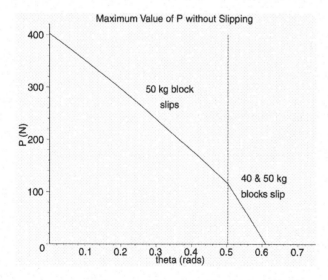

6.2 Problem 6/09 (Friction)

The 30-kg homogeneous cylinder of 400-mm diameter rests against the vertical and inclined surfaces as shown. Calculate the applied clockwise couple M which would cause the cylinder to slip. Also calculate the normal contact forces at A and B. Express your answers in terms of the angle θ and the coefficient of static friction μ_s. (a) Plot the normal forces as a function of θ $(0 \le \theta \le 45°)$ for $\mu = 0.3$, (b) Plot the Moment as a function of θ $(0 \le \theta \le 45°)$ for $\mu = 0.2$, 0.5, and 0.8, (c) Plot the Moment as a function of μ $(0 \le \mu \le 1)$ for $\theta = 15$, 30, and 45°.

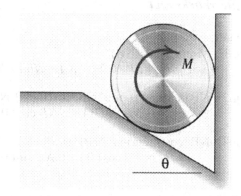

Problem Formulation

First we write the equilibrium equations from the free-body diagram for the cylinder (shown to the right).

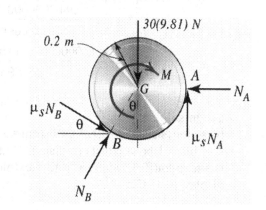

$$\circlearrowleft \Sigma M_A = 0 = M - \mu_s(N_A + N_B)0.2$$

$$\Sigma F_x = 0 = N_B \sin\theta + \mu_s N_B \cos\theta - N_A$$

$$\Sigma F_y = 0 = N_B \cos\theta - \mu_s N_B \sin\theta + \mu_s N_A - 30(9.81)$$

The last two equations can be readily solved simultaneously for N_A and N_B. Substituting these results into the first equation gives an expression that can be solved for M. The results are,

$$N_A = \frac{294.3(\sin\theta + \mu\cos\theta)}{\cos\theta(1+\mu^2)} \qquad N_B = \frac{294.3}{\cos\theta(1+\mu^2)}$$

$$M = \frac{58.86\mu(1+\sin\theta+\mu\cos\theta)}{\cos\theta(1+\mu^2)}$$

Of course, we will let Maple solve the three equations above.

Maple Worksheet

> restart; with(plots):
> eqn1:=M-mu*(NA+NB)*0.2=0;

$$eqn1 := M - 0.2\,\mu\,(NA + NB) = 0$$

> eqn2:=NB*sin(theta)+mu*NB*cos(theta)-NA=0;

$$eqn2 := NB\sin(\theta) + \mu\,NB\cos(\theta) - NA = 0$$

> eqn3:=NB*cos(theta)-mu*NB*sin(theta)-30*9.81+mu*NA=0;

$$eqn3 := NB\cos(\theta) - \mu\,NB\sin(\theta) - 294.30 + \mu\,NA = 0$$

> soln:=solve({eqn1,eqn2,eqn3},{NA,NB,M});

$$soln := \{\, NA = \frac{294.3000000\ (\sin(\theta) + \mu\cos(\theta))}{\cos(\theta)\,(1.+\mu^2)},\ NB = \frac{294.3000000}{\cos(\theta)\,(1.+\mu^2)},$$

$$M = \frac{58.86000000\ \mu\,(\sin(\theta) + \mu\cos(\theta) + 1.)}{\cos(\theta)\,(1.+\mu^2)}\,\}$$

> assign(soln);
> p1:=plot([subs(mu=.3,NA),subs(mu=.3,NB)],theta=0..Pi/4,
color=black,labels=["theta (radians)",""],title="Force (N)"):
> t1:=textplot([0.3,185,"NA"]):t2:=textplot([0.3,300,"NB"]):
> display(p1,t1,t2);

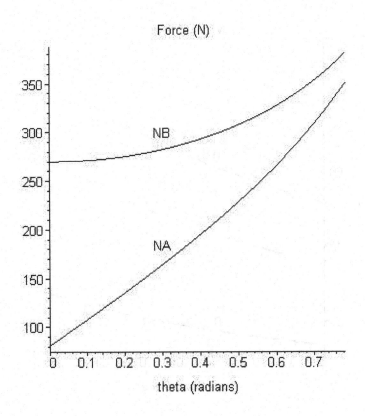

> M1:=subs(mu=.2,M):M2:=subs(mu=.5,M):M3:=subs(mu=0.8,M):

> p2:=plot([M1,M2,M3],theta=0..Pi/4,color=black, labels=["theta (radians)",""],title="Moment (N-m)"):

> t3:=textplot([0.3,20,"0.2"]):t4:=textplot([0.3,47,"0.5"]):t5:=textplot([0.3,65,"0.8"]):

> display(p2,t3,t4,t5);

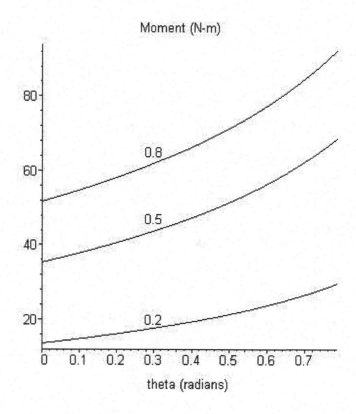

> M4:=subs(theta=15*Pi/180,M):M5:=subs(theta=30*Pi/180,M):M6:=subs(theta=45*Pi/180,M):

> p3:=plot([M4,M5,M6],mu=0..1,color=black,labels=["coefficient of friction",""],title="Moment (N-m)"):

> t6:=textplot([0.9,60,"15"]):t7:=textplot([0.9,82,"30"]):t8:=textplot([0.9,101,"45"]):

> display(p3,t6,t7,t8);

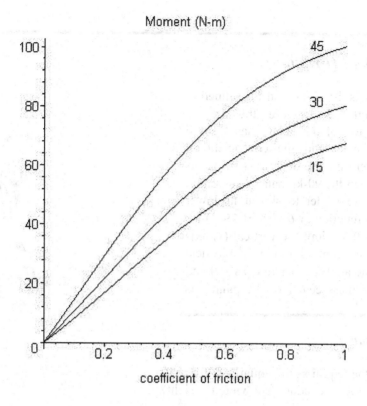

6.3 Problem 6/44 (Friction)

The 15-kg ladder is 4-m long and is inclined at an arbitrary angle θ (ignore the 1.5 m dimension). The top of the ladder has a small roller, and at the ground the coefficient of friction is μ. Determine the distance s (in terms of θ and μ) to which the 90-kg painter can climb without causing the ladder to slip at its lower end A. Plot s as a function of θ ($0 \le \theta \le 90°$) for $\mu = 0.2$, 0.4, and 0.6. Limit the vertical (s) axis to be from 0 to 5-m. Provide a physical explanation for the angles θ where (a) $s = 0$ and (b) $s \ge 4$-m. The mass center of the painter is directly above his feet.

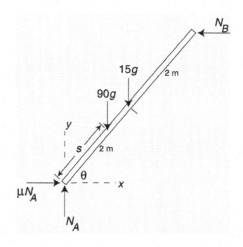

Problem Formulation

First we write the equilibrium equations from the free-body diagram for the ladder (shown to the right).

$$\Sigma F_x = 0 = \mu N_A - N_B$$

$$\Sigma F_y = 0 = N_A - 90g - 15g$$

$$\circlearrowleft \Sigma M_A = 0 = 90g(s\cos\theta) + 15g(2\cos\theta) - N_B(4\sin\theta)$$

The first two equations can be readily solved for $N_B = 105\mu g$. Substituting this result into the third equation gives an expression that can be solved for s.

$$s = \frac{14\mu\sin\theta - \cos\theta}{3\cos\theta}$$

Maple Worksheet

> restart; with(plots):
> NA:=105*g; NB:=mu*NA;

$$NA := 105\,g$$

$$NB := 105\,\mu\,g$$

```
> sumMA:=90*g*s*cos(theta)+15*g*2*cos(theta)-NB*4*sin(theta);
```

$$sumMA := 90\,g\,s\,\cos(\theta) + 30\,g\,\cos(\theta) - 420\,\mu\,g\,\sin(\theta)$$

```
> s:=solve(sumMA,s);
```

$$s := \frac{1}{3}\frac{-\cos(\theta) + 14\,\mu\,\sin(\theta)}{\cos(\theta)}$$

```
> s2:=subs(mu=0.2,s): s4:=subs(mu=0.4,s): s6:=subs(mu=0.6,s):
> t1:=textplot([1.37,3,"0.2"]):
  t2:=textplot([1.15,3,"0.4"]):
  t3:=textplot([0.7,3,"mu=0.6"]):
> p1:=plot([s2,s4,s6],theta=0..Pi/2, y=0..5 ,labels=["theta (rads)","s (m)"],
labeldirections=[HORIZONTAL,VERTICAL]):
> display(p1,t1,t2,t3);
```

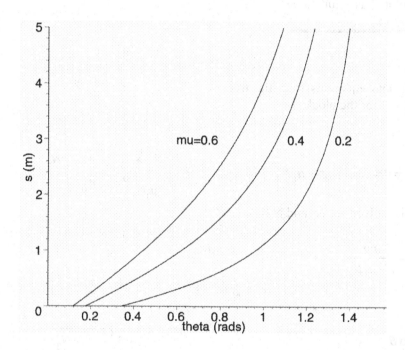

The physical interpretation is as follows. The value of θ for which $s = 0$ corresponds to the orientation where the ladder is on the verge of slipping before the man steps on it. Obviously, the ladder will slip immediately if he does try to step on it. The values of θ for which $s \geq 4$ correspond to orientations where the man can climb to the top of the ladder without having it slip.

6.4 Sample Problem 6/6 (Friction on Wedges)

The horizontal position of the 500-kg rectangular block of concrete is adjusted by the wedge under the action of the force **P**. Let the wedge angle be θ (instead of 5°). Also let μ_1 and μ_2 be the coefficient of static friction at the two wedge surfaces and between the block and the horizontal surface respectively. (a) Derive a general expression for P (the least force required to move the block) in terms of θ, μ_1 and μ_2. (b) For $\mu_2 = 0.6$, plot P as a function of μ_1 for $\theta = 5$, 15, and 25°. (c) For $\theta = 5°$, plot P as a function of μ_1 for $\mu_2 = 0.2, 0.4, 0.6,$ and 0.8.

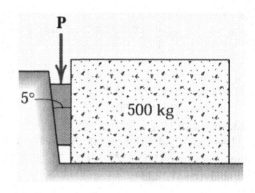

Problem Formulation

(a) First we write the equilibrium equations from the free-body diagrams for the wedge and for the block.

For the Block

$$\Sigma F_x = 0 = N_2 - \mu_2 N_3 \qquad \Sigma F_y = 0 = N_3 - mg - \mu_1 N_2$$

These two equations are readily solved for N_2 and N_3.

$$N_2 = \frac{\mu_2 mg}{1 - \mu_1 \mu_2} \qquad\qquad N_3 = \frac{mg}{1 - \mu_1 \mu_2}$$

For the Wedge

$$\Sigma F_x = 0 = N_1 \cos\theta - \mu_1 N_1 \sin\theta - N_2$$
$$\Sigma F_y = 0 = N_1 \sin\theta + \mu_1 N_1 \cos\theta + \mu_1 N_2 - P$$

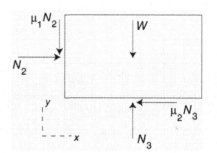

After substituting for N_2 we solve the first equation for N_1. Substituting this result into the second yields an expression for P. The result is also obtained, in somewhat different form, by symbolic algebra in the worksheet below.

$$P = \frac{\mu_2 mg\left(2\mu_1 + \left(1 - \mu_1^2\right)\tan\theta\right)}{\left(1 - \mu_1 \tan\theta\right)\left(1 - \mu_1 \mu_2\right)}$$

This result is used to generate the plots for parts **(b)** and **(c)** in the worksheet below.

Maple Worksheet

> restart; with(plots):

From the free-body diagram for the block

> N2:=mu[2]*m*g/(1-mu[1]*mu[2]);
$$N2 := \frac{\mu_2 \, m \, g}{1 - \mu_1 \, \mu_2}$$

> N3:=m*g/(1-mu[1]*mu[2]);
$$N3 := \frac{m \, g}{1 - \mu_1 \, \mu_2}$$

For the wedge

> sumFx:=N1*cos(theta)-mu[1]*N1*sin(theta)-N2=0;
$$sumFx := N1 \cos(\theta) - \mu_1 \, N1 \sin(\theta) - \frac{\mu_2 \, m \, g}{1 - \mu_1 \, \mu_2} = 0$$

> sumFy:=N1*sin(theta)+mu[1]*N1*cos(theta)+mu[1]*N2-P=0;
$$sumFy := N1 \sin(\theta) + \mu_1 \, N1 \cos(\theta) + \frac{\mu_1 \, \mu_2 \, m \, g}{1 - \mu_1 \, \mu_2} - P = 0$$

> soln:=solve({sumFx,sumFy},{N1,P});
$$soln := \left\{ P = -\frac{\mu_2 \, m \, g \, (\mu_1{}^2 \sin(\theta) - 2 \, \mu_1 \cos(\theta) - \sin(\theta))}{\cos(\theta) - \cos(\theta) \, \mu_1 \, \mu_2 - \mu_1 \sin(\theta) + \mu_1{}^2 \sin(\theta) \, \mu_2}, \right.$$
$$\left. N1 = \frac{\mu_2 \, m \, g}{\cos(\theta) - \cos(\theta) \, \mu_1 \, \mu_2 - \mu_1 \sin(\theta) + \mu_1{}^2 \sin(\theta) \, \mu_2} \right\}$$

> assign(soln);
> m:=500: g:=9.81:

Part (b)
> mu[2]:=0.6:
> P_5:=subs(theta=5*Pi/180,P)/1000; # converting to kN
> P_15:=subs(theta=15*Pi/180,P)/1000:
  P_25:=subs(theta=25*Pi/180,P)/1000:
> p1:=plot([P_5,P_15,P_25],mu[1]=0..0.8,labels=[`mu_1`,`P (kN)`],
labeldirections=[HORIZONTAL,VERTICAL]):

```
> t1:=textplot([0.8,10.4,`5`]):
> t2:=textplot([0.8,12.75,`15`]):
> t3:=textplot([0.8,16.5,`25`]):
> display(p1,t1,t2,t3);
```

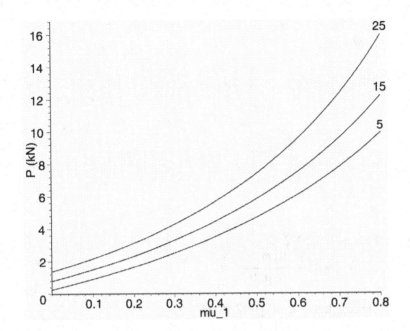

Part (c)
```
> theta:=5*Pi/180: # converts to radians
> unassign('mu[2]');
> P_2:=subs(mu[2]=0.2,P)/1000; # converts to kN
> P_4:=subs(mu[2]=0.4,P)/1000:
  P_6:=subs(mu[2]=0.6,P)/1000:
  P_8:=subs(mu[2]=0.8,P)/1000:
> p2:=plot([P_2,P_4,P_6,P_8],mu[1]=0..0.8,labels=[`mu_1`,`P (kN)`],
labeldirections=[HORIZONTAL,VERTICAL]):
> t4:=textplot([0.8,2.7,`0.2`]): t5:=textplot([0.8,5.7,`0.4`]):
> t6:=textplot([0.8,10.7,`0.6`]): t7:=textplot([0.8,19.7,`0.8`]):
> display(p2,t4,t5,t6,t7);
```

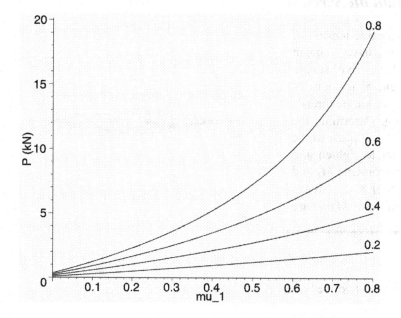

6.5 Problem 6/60 (Friction on Screws)

The large turnbuckle supports a cable tension of 10,000 lb. The 1.25 in. screws have a mean diameter of 1.150 in. and have five square threads per inch. The coefficient of friction for the threads is μ. Both screws have single threads and are prevented from turning. Determine the moments M_T and M_L that must be applied to the body of the turnbuckle in order to tighten and loosen it respectively. Plot the moments M_T and M_L as functions of μ for $0 \leq \mu \leq 1$. Give a physical explanation for any values of M that are less than zero.

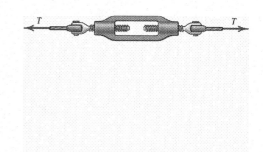

Problem Formulation

From equations 6/3 and 6/3b in your text we have,

$$M_T = 2Tr \tan(\phi + \alpha) \qquad M_L = 2Tr \tan(\phi - \alpha)$$

where T = 10,000 lb, the lead L = 1/5 in./rev. and the mean radius r = 1.15/2 = 0.575 in. Also,

$$\alpha = \tan^{-1} \frac{L}{2\pi r} \qquad \phi = \tan^{-1} \mu$$

Substitution will give M_T and M_L explicitly as a function of μ. We will let the computer substitute for us.

Maple Worksheet

```
> restart; with(plots):
> M_T:=2*T*r*tan(phi+alpha);
            M_T := 2 T r tan(φ + α)

> M_L:=2*T*r*tan(phi-alpha);
            M_L := 2 T r tan(φ − α)

> alpha:=arctan(L/2/Pi/r);
            α := arctan( 1  L  )
                       ( 2 π r )

> phi:=arctan(mu);
```

$$\phi := \arctan(\mu)$$

```
> T:=10000: L:= 1/5: r:=1.150/2:
>
> t1:=textplot([.8,12000,"to tighten"]):
> t2:=textplot([.8,7000,"to loosen"]):
> p:=plot([M_T,M_L],mu=0..1, labels=["mu","Moment (lb-in)"],
labeldirections=[HORIZONTAL,VERTICAL]):
> display(p,t1,t2);
```

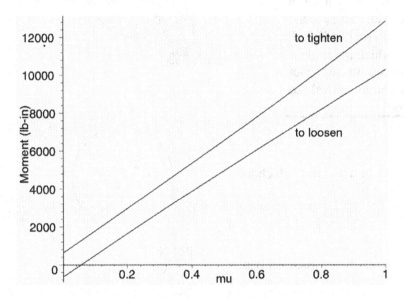

Note that the predicted values for M_L are negative for small μ. What is the physical explanation for this? You may recall that the equation for M_L derived in your text assumes that $\phi > \alpha$. When $\phi = \alpha$, the turnbuckle will be on the verge of loosening without an external moment applied to its body. The condition $\phi = \alpha$ gives, from the equations above,

$$\mu = \frac{L}{2\pi r} = \frac{1/5}{2\pi(0.575)} = 0.00553$$

This is the value of μ for which $M_L = 0$ in the plot above. We conclude then that the turnbuckle will loosen without any external moment being applied whenever $\mu < 0.00553$.

6.6 Sample Problem 6/9 (Flexible Belts)

A flexible cable which supports the 100-kg load is passed over a circular drum and subjected to a force P to maintain equilibrium. The coefficient of static friction between the cable and the fixed drum is μ. (a) For $\alpha = 0$, determine the maximum and minimum values P may have in order not to raise or lower the load. Plot P_{max} and P_{min} versus μ for $0 \leq \mu \leq 1$. (b) For $P = 500$ N, determine the minimum value which the angle α may have before the load begins to slip. Plot α_{min} versus μ for $0 \leq \mu \leq 1$. Limit to vertical (α) axis to be between -60 and $360°$.

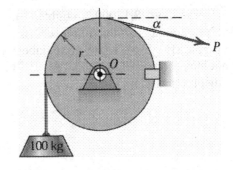

Problem Formulation

Here we will use Equation 6/7 in your text which is repeated below for convenience.

$$T_2 = T_1 e^{\mu\beta}$$

Recall that in deriving this formula it was assumed that $T_2 > T_1$.

(a) With $\alpha = 0$ the contact angle is $\beta = \pi/2$ rad. For impending upward motion of the load we have $T_2 = P_{max}$ and $T_1 = 981$ N. Thus

$$P_{max} = 981 e^{\mu\pi/2}$$

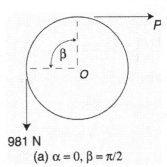

(a) $\alpha = 0, \beta = \pi/2$

For impending downward motion of the load we have $T_2 = 981$ N and $T_1 = P_{min}$.

$$981 = P_{min} e^{\mu\pi/2} \quad \text{or} \quad P_{min} = 981 e^{-\mu\pi/2}$$

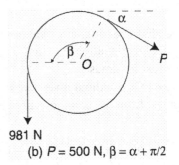

(b) $P = 500$ N, $\beta = \alpha + \pi/2$

(b) With $P = 500$ N we have $\beta = \pi/2 + \alpha$, $T_2 = 981$ N and $T_1 = P = 500$ N. From Equation 6/7 we have,

$$981/500 = e^{\mu(\pi/2+\alpha)}$$

Taking the natural log of both sides of the above equation and solving for α gives,

$$\alpha = \frac{\ln(981/500)}{\mu} - \frac{\pi}{2}$$

Maple Worksheet

> restart; with(plots):
> Pmax:=981*exp(mu*Pi/2);

$$Pmax := 981\, e^{(1/2\,\mu\,\pi)}$$

> Pmin:=981*exp(-mu*Pi/2);

$$Pmin := 981\, e^{(-1/2\,\mu\,\pi)}$$

> t1:=textplot([.8,4,"Pmax"]):
> t2:=textplot([.8,.5,"Pmin"]):
> p1:=plot([Pmax/1000,Pmin/1000],mu=0..1, labels=["coefficient of friction", "P(kN)"], labeldirections=[HORIZONTAL,VERTICAL]):
> display(p1,t1,t2);

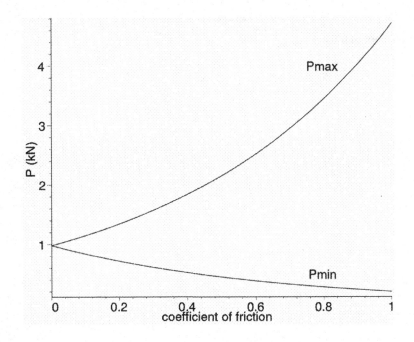

> alpha:=log(981/500)/mu-Pi/2;

$$\alpha := \frac{\ln\left(\dfrac{981}{500}\right)}{\mu} - \frac{1}{2}\pi$$

> alpha:=alpha*180/Pi: # converts from radians to degrees
> plot(alpha, mu=0..1,y=-60..360,labels=["coefficient of friction","alpha
(deg)"],labeldirections=[HORIZONTAL,VERTICAL]);

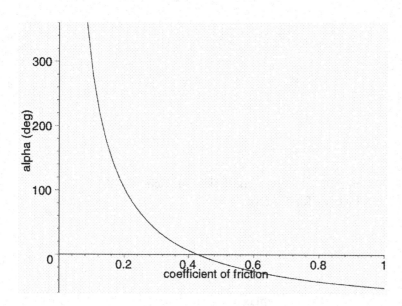

VIRTUAL WORK

7

This chapter considers the application of the principle of virtual work to the equilibrium and stability analysis of engineering structures. Problem 6.1 looks at the compressive force developed in a hydraulic cylinder that is part of a mechanism for elevating automobiles. The results of a parametric study suggest an obvious improvement in the design of the system. The formulation of problem 7.2 results in an equation that cannot be solved exactly. When this occurs one generally has to make a choice between a graphical or numerical solution. This problem illustrates in some detail a general graphical approach that is very useful in some situations. For purposes of comparison, a numerical solution is also obtained with Maple's *fsolve* function. Problem 7.3 takes a look at a problem with multiple equilibrium states and evaluates these in terms of their stability.

Following is a summary of the Maple commands used in this chapter.

Table 7.1 Maple Commands used in Chapter 7

Maple Command	Problem
fsolve	7.2
diff	7.3
display	7.1, 7.2, 7.3
plot (normal)	7.1, 7.2, 7.3
solve	7.3
subs	7.1, 7.2, 7.3
textplot	7.1, 7.2, 7.3

7.1 Problem 7/30 (Virtual Work, Equilibrium)

Express the compression C in the hydraulic cylinder of the car hoist in terms of the angle θ. The mass of the hoist is negligible compared to the mass m of the vehicle. (a) Plot the non-dimensional compression C/mg as a function of θ for three non-dimensional length ratios; $b/L =$ 0.1, 0.5, and 0.9. Let θ range between 0 and 90° and limit the vertical scale to between 0 and 3. (b) Plot C/mg as a function of b/L for several values of θ letting b/L vary between 0 and 2. From your results, see if you can improve the design of the hoist system by re-positioning the hydraulic cylinder.

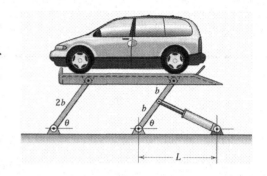

Problem Formulation

The active force diagram for the system is shown to the right. Let the length of the cylinder (AB in the diagram) be a so that the virtual work done by C is $C\delta a$. The principle of virtual work applied to this system gives,

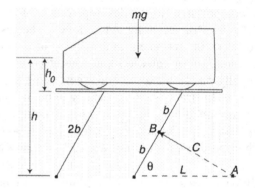

$$\delta U = 0 = C\delta a - mg\delta h$$

Now we have to do some geometry to relate the virtual displacements δa and δh to the angle θ. From the diagram we can write a^2 as follows.

$$a^2 = (b\sin\theta)^2 + (L - b\cos\theta)^2$$

Note that it is not actually necessary to solve for a in order to determine the virtual displacement δa. All you need is an expression relating a and θ. In the present case it is somewhat easier to find the variation of the expression on the right hand side of the equation above than it would be to find the variation of the square root of that expression. Taking the variation of the above equation yields,

$$2a\delta a = 2b\sin\theta(b\cos\theta\delta\theta) + 2(L - b\cos\theta)(b\sin\theta\delta\theta)$$

Solving for δa and simplifying yields,

$$\delta a = \frac{Lb}{a}\sin\theta\delta\theta$$

where $\quad a = \sqrt{(b\sin\theta)^2 + (L - b\cos\theta)^2} = L\sqrt{1 + \left(\frac{b}{L}\right)^2 - 2\frac{b}{L}\cos\theta}$

Now we need to find a relationship between δh and θ.

$$h = 2b\sin\theta + h_0 \qquad \delta h = 2b\cos\theta\delta\theta$$

Now we substitute the virtual displacements δa and δh into the virtual work equation above.

$$C\left(\frac{Lb}{a}\sin\theta\right)\delta\theta - mg(2b\cos\theta)\delta\theta = 0$$

$$C = \frac{2mga}{L}\cot\theta$$

Solving for the non-dimensional compression C/mg and substituting for a yields,

$$\frac{C}{mg} = 2\cot\theta\sqrt{1 + \left(\frac{b}{L}\right)^2 - 2\frac{b}{L}\cos\theta}$$

Before generating the required plots it might be useful to introduce some non-dimensional parameters. Letting $C' = C/mg$ and $\eta = b/L$ we can re-write the above equation as

$$C' = 2\cot\theta\sqrt{1 + \eta^2 - 2\eta\cos\theta}$$

Now we need to plot C' as a function of θ for $\eta = 0.1$, 0.5, and 0.9.

*** Maple Worksheet***

```
> restart; with(plots):
> C:=2*cot(theta)*sqrt(1+eta^2-2*eta*cos(theta));
        C := 2 cot(θ) √(1 + η² - 2 η cos(θ))
```

Part (a)

```
> C1:=subs(eta=0.1,C):
```

```
> C5:=subs(eta=0.5,C): C9:=subs(eta=0.9,C):
> t1:=textplot([.8,2.5,"b/L = 0.1"]):
> t5:=textplot([.52,2.5,"0.5"]): t9:=textplot([.2,2.5,"0.9"]):
> p1:=plot([C1,C5,C9],theta=0..Pi/2,y=0..3,labels=["theta (rads)","C/mg"],
labeldirections=[HORIZONTAL,VERTICAL]):
> display(p1,t1,t5,t9);
```

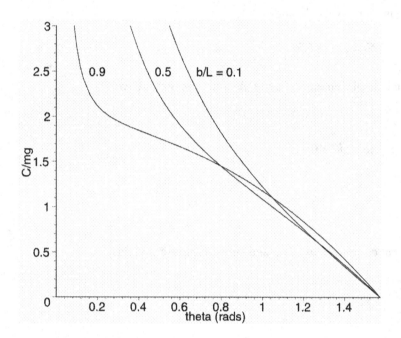

Part (b)

```
> C10:=subs(theta=10*Pi/180,C): C30:=subs(theta=30*Pi/180,C):
> C50:=subs(theta=50*Pi/180,C): C70:=subs(theta=70*Pi/180,C):
> t10:=textplot([1.07,2.55,"10"]): t30:=textplot([1.25,2,"30"]):
> t50:=textplot([1.4,1.6,"50"]): t70:=textplot([1.6,1,"70"]):
> p2:=plot([C10,C30,C50,C70], eta=0..2,y=0..3, labels=["b/L","C/mg"],
labeldirections=[HORIZONTAL,VERTICAL]):
> display(p2,t10,t30,t50,t70);
```

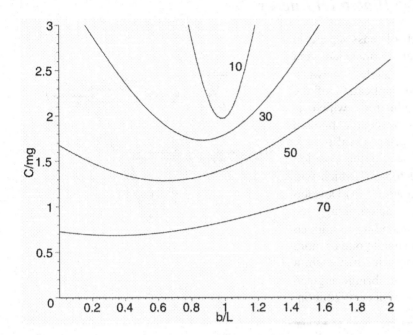

The first diagram (part (a)) shows that the present design has some real problems for small angles θ. To appreciate this, remember that we are plotting the non-dimensional force. If the non-dimensional force equals 3, for example, the compressive force in the cylinder will be three times the weight of the vehicle. The main intent of the diagram for part (b) is to show that changing the design by fine-tuning b/L will not really remove the real problem.

So, what is the physical reason for the compressive force approaching infinity as θ approaches zero? Look again at the active force diagram and try to visualize how the orientation of C changes as θ gets smaller. As θ gets smaller, the vertical component of C gets smaller, making it harder and harder for the cylinder to support the weight of the vehicle. Without changing the overall design too much, the most obvious thing to do is to recess the cylinder somewhat so that it never approaches a horizontal orientation. In other words, move point A (the base of the cylinder) vertically downward. This is illustrated by the modified drawing to the right. Note that, with this arrangement, the vertical component of C will not approach zero at small θ.

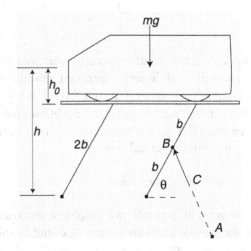

7.2 Sample Problem 7/5 (Potential Energy)

The two uniform links, each of mass m, are in the vertical plane and are connected and constrained as shown. As the angle θ between the links increases with the application of the horizontal force P, the light rod, which is connected at A and passes through a pivoted collar at B, compresses the spring of stiffness k. If the spring is uncompressed in the position where $\theta = 0$, determine the force P which will produce equilibrium at the angle θ. (a) Develop a graphical approach for determining the equilibrium value of θ corresponding to a given force P. Use a computer to generate one or more plots that illustrate your approach. (b) Obtain a numerical solution for the equilibrium angle θ for the case $P = 100$ lb, $mg = 50$ lb, $k = 20$ lb/in, and $b = 5$ in.

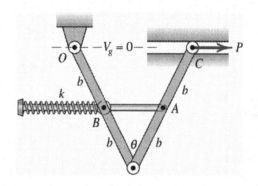

Problem Formulation

The first part of this problem (determining the force P which will produce equilibrium at the angle θ) is identical to the sample problem in your text. Therefore, we start with the equation obtained in the sample problem. Be sure that you understand how this equation was derived.

$$P = kb \sin \frac{\theta}{2} + \frac{1}{2} mg \tan \frac{\theta}{2}$$

This equation is ideally suited to situations where you would like to find the force P that will result in some specified equilibrium angle θ. Here we are interested in the opposite situation. Given a force P we would like to calculate the equilibrium value of θ. Unfortunately, the equation above cannot be inverted analytically to give θ explicitly as a function of P. In situations like this you should try either a graphical or numerical solution.

(a) Graphical Approach

Here we will illustrate two graphical approaches. The first is suggested at the end of the sample problem in your text and is suitable for solving specific problems. The second approach is more general and is suitable for certain types of design situations.

(a1) *Specific Graphical Approach*

The specific approach is suitable for situations where P, b, m and k are all specified. For example, consider the case where you want the equilibrium angle θ for only one situation. For convenience, we'll take the specific case defined in part (b).

$$P = 100 \text{ lb}, mg = 50 \text{ lb}, k = 20 \text{ lb/in, and } b = 5 \text{ in.}$$

Here we forget for the moment that P is given and plot, from the expression above, P as a function of θ. Superimposed on this plot would be a horizontal line at $P = 100$. This horizontal line can either be drawn at the same time the plot is generated or by hand after the plot has been printed. After the plot has been printed you would next find the intersection between the horizontal line and the curve. Since the curve represents the solution for all θ, the intersection will be the solution for the specific case where $P = 100$ lb. The equilibrium value of θ can thus be found by dropping a vertical line from the point of intersection to the θ axis. The procedure is illustrated in the figure to the right. From this plot we can estimate the equilibrium angle to be about $\theta = 95°$. For purposes of comparison, the numerical approach in part (b) gives $\theta = 94.05°$.

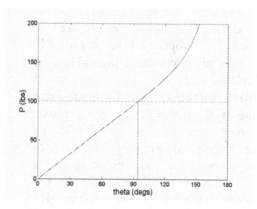

(a2) *General Graphical Approach*

What we have in mind here are certain types of design situations where specific values of parameters have not yet been decided. Here we want to develop a graphical approach that would not require us to run to a computer each time a specific case is considered. This is what is meant by a general graphical approach. The effectiveness and generality of graphical solutions can be significantly increased by first reducing, as much as possible, the number of parameters that appear in the equation. In the present case we can reduce the number of parameters from three (k, b, and m) to one by dividing both sides of the equation by the weight mg. Since P, kb and mg all have units of force, this operation results in a non-dimensional equation relating a non-dimensional force $P' = P/mg$ to the equilibrium angle θ. The only parameter remaining is the non-dimensional quantity $kb/mg = \beta$.

$$P' = \frac{P}{mg} = \beta \sin\frac{\theta}{2} + \frac{1}{2}\tan\frac{\theta}{2} \quad \text{where} \quad \beta = \frac{kb}{mg}.$$

The plot to the right shows P' versus θ for several, evenly spaced, values of β. For a specific problem you would be asked to find θ given P, k, b, and m. One would first determine P' and β and then construct a horizontal line at the calculated value of P' and find where that line intersects a curve at the appropriate value of β. Dropping a vertical line from this intersection gives the angle θ.

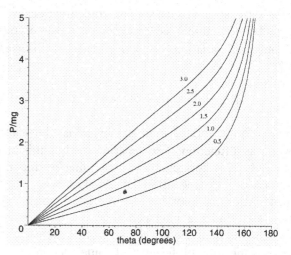

Of course, if the graph is prepared in advance, as proposed here, it probably will not have a curve for the value of β calculated. In this case you could generate another curve at the appropriate value of β, but this defeats our purpose. Another approach is to obtain an approximate answer by means of interpolation. To illustrate, consider a case where $P = 220$ lb, $mg = 100$ lb, $k = 29$ lb/in and $b = 6$ in. These values give $P' = 2.2$ and $\beta = 1.74$. The second plot to the right illustrates the interpolation procedure. Start with the horizontal line at $P' = 2.2$ and then follow this line to where it reaches a point about halfway between the curves for $\beta = 1.5$ and 2. Dropping a vertical line from this point gives a value for the equilibrium angle θ of about 112°.

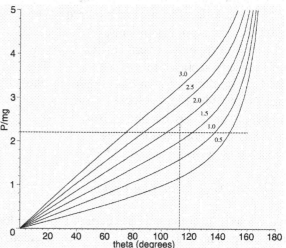

The worksheet below will show how to generate plots such as those shown above. Of course, the construction of vertical lines and interpolation are operations that are performed after the curves are generated and printed.

(b) The numerical solution is given in the worksheet below.

Maple Worksheet

> restart; with(plots):with(plottools):

Plot for part (a1)

> Pa:=k*b*sin(theta/2)+1/2*mg*tan(theta/2);

$$Pa := k\, b\, \sin\!\left(\frac{1}{2}\,\theta\right) + \frac{1}{2}\, mg\, \tan\!\left(\frac{1}{2}\,\theta\right)$$

The following re-writes the force so that it will be a function of x where x is the angle theta expressed in degrees.

> Pa:=subs(theta=x*Pi/180,Pa); #

$$Pa := k\, b\, \sin\!\left(\frac{1}{360}\,x\,\pi\right) + \frac{1}{2}\, mg\, \tan\!\left(\frac{1}{360}\,x\,\pi\right)$$

> P100:=100: k:=20: mg:=50: b:=5:
> plot([Pa,P100],x=0..180,y=0..200, labels=["theta (degs)","P (lbs)"], labeldirections=[HORIZONTAL,VERTICAL]);

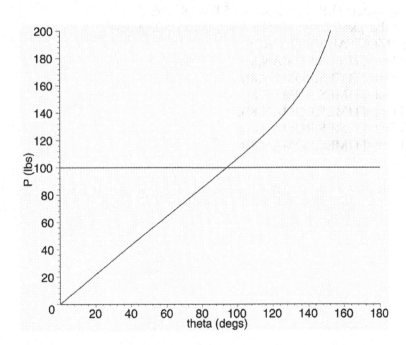

Part (b)

It is convenient to do part (b) here since it involves the parameters already defined above. One of the points here is, of course, that it is very difficult to beat the numerical approach for accuracy and ease of application. We find the result with a single command.

```
> fsolve(Pa=P100,x);
                94.05206478
```

Plot for part (a2)

```
> Pprime:=beta*sin(theta/2)+1/2*tan(theta/2);
```
$$Pprime := \beta \, \sin\left(\frac{1}{2}\,\theta\right) + \frac{1}{2}\,\tan\left(\frac{1}{2}\,\theta\right)$$

```
> Pd:=subs(theta=x*Pi/180,Pprime); # converting to degrees as in (a1) above
```
$$Pd := \beta \, \sin\left(\frac{1}{360}\,x\,\pi\right) + \frac{1}{2}\,\tan\left(\frac{1}{360}\,x\,\pi\right)$$

```
>p1:=plot([subs(beta=0.5,Pd),subs(beta=1,Pd),subs(beta=1.5,Pd),subs(beta=2,Pd
),subs(beta=2.5,Pd),subs(beta=3,Pd)],x=0..180,y=0..5,labels=[`theta    (degrees)`
,`P/mg`], labeldirections=[HORIZONTAL,VERTICAL]):
> t1:=textplot([140,2.05,`0.5`],font=[TIMES,ROMAN,8]):
> t2:=textplot([135,2.35,`1.0`],font=[TIMES,ROMAN,8]):
> t3:=textplot([130,2.65,`1.5`],font=[TIMES,ROMAN,8]):
> t4:=textplot([125,2.95,`2.0`],font=[TIMES,ROMAN,8]):
> t5:=textplot([120,3.25,`2.5`],font=[TIMES,ROMAN,8]):
> t6:=textplot([115,3.55,`3.0`],font=[TIMES,ROMAN,8]):
> p2:=display(p1,t1,t2,t3,t4,t5,t6):
> display(p2);
```

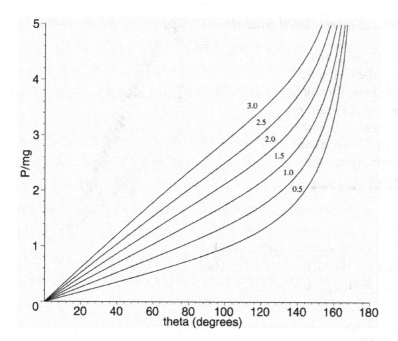

7.3 Problem 7/79 (Potential Energy and Stability)

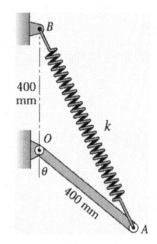

The uniform link OA has a mass of 20 kg and is supported in the vertical plane by the spring AB whose unstretched length is 400 mm. Plot the total potential energy V as a function of θ from $\theta = -15°$ to $\theta = 120°$. Consider two cases, $k = 100$ and 900 N/m. Determine all equilibrium positions (angles θ) and their stability for each case. Take $Vg = 0$ on a level through O.

Problem Formulation

Imagine dividing triangle OAB into two identical right triangles to note that half of the length AB is $0.4\sin((180 - \theta)/2)$. Also note the trig identity $\sin((180 - \theta)/2) = \cos(\theta/2)$. Thus,

$$AB = 2(0.4)\sin\left(\frac{180 - \theta}{2}\right) = 0.8\cos\frac{\theta}{2}$$

Since the unstretched length of the spring is 0.4 m, the elastic potential energy for the springs is,

$$V_e = \frac{1}{2}k(AB - 0.4)^2 = \frac{1}{2}k\left(0.8\cos\frac{\theta}{2} - .4\right)^2$$

Taking $V_g = 0$ on the horizontal plane through O we obtain the gravitational potential energy of the system as,

$$V_g = 20(9.81)(-0.2\cos\theta) = -39.24\cos\theta$$

Thus, the total potential energy and its derivative are,

$$V = V_g + V_e = \frac{1}{2}k\left(0.8\cos\frac{\theta}{2} - 0.4\right)^2 - 39.24\cos\theta$$

$$\frac{dV}{d\theta} = 39.24\sin\theta - 0.4k\left(0.8\cos\frac{\theta}{2} - 0.4\right)\sin\frac{\theta}{2}$$

The equilibrium positions and their stability will be determined below.

Maple Worksheet

```
> restart; with(plots):
> AB:=2*0.4*sin(Pi/2-theta/2);
```

$$AB := 0.8 \ \cos\left(\frac{\theta}{2}\right)$$

```
> Ve:=1/2*k*(AB-0.4)^2;
```

$$Ve := \frac{1}{2} k \left(0.8 \ \cos\left(\frac{\theta}{2}\right) - 0.4 \right)^{2}$$

```
> Vg:=-20*9.81*0.2*cos(theta);
```

$$Vg := -39.240 \ \cos(\theta)$$

```
> V:=Vg+Ve;
```

$$V := -39.240 \ \cos(\theta) + \frac{1}{2} k \left(0.8 \ \cos\left(\frac{\theta}{2}\right) - 0.4 \right)^{2}$$

```
> t1:=textplot([1,27,"k = 900 N/m"]): t2:=textplot([1,-24,"k = 100 N/m"]):
> p1:=plot([subs(k=100,V),subs(k=900,V)],theta=-15*Pi/180..120*Pi/180, color=black,
labels=["theta (rads)","V (J)"]):
> display(p1,t1,t2);
```

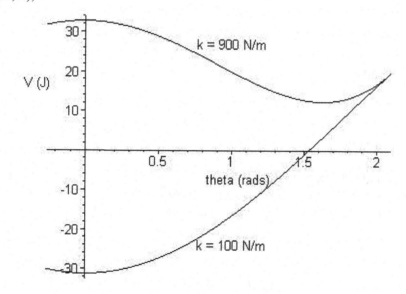

Equilibrium positions and their stability.

Recall from your text that stable equilibrium occurs at positions where the total potential energy of the system is a minimum. Unstable equilibrium, on the other hand, occurs at positions where the total potential energy of the system is a maximum. From this general principle we see that the spring with stiffness 100 N/m has only one equilibrium position at $\theta = 0$ and it is stable. The spring with stiffness 900 N/m has two equilibrium positions, an unstable one at $\theta = 0$ and a stable one just beyond 1.5 radians. This second position can be determined more precisely by finding the angles for which $dV/d\theta = 0$. This is carried out in the following.

> dV:=diff(V,theta);

$$dV := 39.240 \, \sin(\theta) - 0.4000000000 \; k \left(0.8 \, \cos\left(\frac{\theta}{2}\right) - 0.4 \right) \sin\left(\frac{\theta}{2}\right)$$

> k:=900;

$$k := 900$$

> solve(dV=0,theta);

$$6.283185307 \;, 0., -1.626102561 \;, 1.626102561$$

Maple has found four solutions, but only the second and fourth are in our range. The second solution (0) is unstable while the fourth (1.626 radians) is stable

Summary:

$k = 100$ N/m

$\qquad \theta = 0$ Stable

$k = 900$ N/m

$\qquad \theta = 0$ Unstable

$\qquad \theta = 1.626$ rads (93.18°)   Stable